MOBILE CRANE SUPPORT HANDBOOK

MOBILE CRANE SUPPORT HANDBOOK

Second Edition

David Duerr, P.E.

2DM Associates, Inc.
Houston, Texas

LEVARE PRESS

This book is printed on acid-free paper. ∞

Mobile Crane Support Handbook

ISBN: 978-0-578-42911-3

10   9   8   7   6   5   4   3   2   1

Contents

About the Author.. ix

Preface to the Second Edition xi

Preface to the First Edition xv

1   Crane Loads 1

1.1   Loads from the Crane 1
     1.1.1   Reactions from the Superstructure 1
     1.1.2   Reactions from the Lower 4
     1.1.3   The Total Lifted Load.. 5
     1.1.4   Where the Challenge Lies 6
1.2   Crane Support Loads 6
     1.2.1   Outrigger Loads.. 8
     1.2.2   Crawler Track Bearing Pressures 12
     1.2.3   Methods for Reducing Track Bearing Pressures 24
     1.2.4   Summary Comments on Crane Support Load
             Calculations 29
1.3   Software for Crane Load Calculations 29
     1.3.1   Crane Manufacturer Web-Based Calculators 30
     1.3.2   Crane Manufacturer Locally Installed Software 30
     1.3.3   Independently Produced Software Products................. 31
1.4   Support Load Patterns 31
     1.4.1   Crawler Track Bearing Pressure Patterns.................... 31
     1.4.2   Truck Crane Outrigger Load Patterns 32
1.5   Variable Load Effects 33
     1.5.1   Effect of Change in Boom Tip Load 34
     1.5.2   Vertical Impact 36
     1.5.3   Horizontal Dynamic Forces... 36
1.6   Application Summary.. 39

2

Strength of the Supporting Surface...41

2.1   Geotechnical Exploration 41
2.2   Soil Properties 45
    2.2.1   Soil Bearing Capacity.................................... 45
    2.2.2   Use of Building Code Values 47
    2.2.3   Elastic Properties of Soil 48
    2.2.4   Other Soil Properties 52
    2.2.5   Bearing Capacity of Frozen Soil 53
2.3   Addressing Low Soil Bearing Capacity 54
2.4   Structures Affected by Surface Loads 56
    2.4.1   Crane Loads Acting on Subsurface Structures............ 57
    2.4.2   Strength of Buried Pipes 60
    2.4.3   Strength of Underground Utilities................. 73
    2.4.4   Bearing Pressure Concentrations.................. 74
    2.4.5   Retaining Walls and Bulkheads.................... 75
    2.4.6   Embankments 78
2.5   Cranes Supported on Structures 78
    2.5.1   Pavement and Slabs on Grade...................... 79
    2.5.2   Framed Structures... 80
    2.5.3   Applicable Standards..................................... 82
2.6   Application Summary... 83
2.7   References 84

3

Crane Mat Strength and Stiffness 87

3.1   Timber Mat Dimensions and Condition Survey........ 87
    3.1.1   Methodology 88
    3.1.2   Survey Results... 89
3.2   Timber Species Survey 92
3.3   Standards for Timber Properties 94
    3.3.1   ASTM Standards 94
    3.3.2   Industry Standards and Guides...................... 95
    3.3.3   Application of the Timber Standards - Defects........... 96
    3.3.4   Application of the Timber Standards - Other
             Adjustments.. 103
3.4   Tie Rods in Timber Crane Mats.............................. 104
3.5   Strength and Stiffness Design Values for Timber Mats.......... 106
    3.5.1   Allowable Stresses Based on NDS (AWC 2016) –
             Common Species..................................... 106

3.5.2　Allowable Stresses Based on the NDS (AWC 2016) Method – Tropical Hardwoods 111

3.5.3　Allowable Stresses Based on the Timber Survey 115

3.5.4　Limit State Stresses for the Common Species 118

3.5.5　Limit State Stresses for Tropical Hardwoods 120

3.5.6　Lower Strength Timber Species 120

3.6　Fabricated Steel Mat Properties 122

3.7　Design Values for Other Materials 124

3.8　Plywood Used Under Cranes 124

3.9　Application Summary 125

3.10　Acknowledgements 126

3.11　References 126

4   Crane Mat Behavior 129

4.1　Past Practice of Crane Mat Analysis 129

4.1.1　Mat Length Based on Ground Bearing Capacity 130

4.1.2　Mat Length Based on Mat Strength 133

4.1.3　Comments on These Design Methods 135

4.2　Balanced Mat Analysis Method 137

4.2.1　Effective Bearing Area Calculation Method 137

4.2.2　Performance of the Balanced Mat Analysis Method .. 142

4.2.3　Crane Mat Stiffness 148

4.2.4　Comments on Support Deflection 150

4.2.5　Eccentrically Loaded Mats 151

4.2.6　Significance of the Modulus of Subgrade Reaction ... 154

4.3　Mats of Other Materials 154

4.3.1　Structural Steel Mats 155

4.3.2　Synthetic Mats 155

4.3.3　Steel Plates 158

4.4　Alternate Mat Arrangements 161

4.4.1　Multiple Layers of Mats – Mats Parallel 161

4.4.2　Multiple Layers of Mats – Mats Perpendicular – Case 1 163

4.4.3　Multiple Layers of Mats – Mats Perpendicular – Case 2 169

4.4.4　Mats Continuous Under Both Crawler Tracks 170

4.4.5　Steel Plate or Synthetic Pad over Timber Mats 172

4.4.6　Laminated Wood Mats 176

4.5　Application Summary 177

4.6　References 179

5

Mobile Crane Support Design 181

5.1	Regulatory Language... 181
5.2	Required Information 183
	5.2.1	Crane Support Loads................................... 184
	5.2.2	Supporting Surface Strength 186
5.3	Crane Mat Design – Bearing on Soil............................ 189
	5.3.1	Crane Mats Under Outriggers 190
	5.3.2	Crane Mats Under Crawler Tracks................ 191
5.4	Crane Mat Design – Bearing on Structures or Pavement........ 201
	5.4.1	Crane Mats on Slabs or Pavement............. 201
	5.4.2	Crane Mats on Structures 203
	5.4.3	Applicable Standards.................................. 204
5.5	Crane Support While Traveling............................ 205
5.6	Rules of Thumb – Do they Work?............................ 206
	5.6.1	Crawler Cranes.. 207
	5.6.2	Outrigger-Supported Cranes...................... 208
5.7	Notes on Practical Applications.............................. 213
5.8	Inspection of Crane Mats....................................... 214
5.9	Applications to Other Types of Equipment 215
5.10	Application Summary – A Personal Note.................. 216
5.11	References 216

Appendix 1 – Glossary of Specialized Terms................................. 219

Appendix 2 – USCU / SI Conversion Factors 227

Appendix 3 – Rounding Conventions... 231

Appendix 4 – Notation.. 235

Index 243

About the Author

David Duerr, P.E. is president of 2DM Associates, Inc., his consulting engineering practice in Houston, Texas. Mr. Duerr is a specialist in engineering for heavy lifting and transportation and began working in the field in 1974. He has been actively researching the behavior of crane mats and developing new methods of mat design since the late 1990s and has presented lectures on the subject at construction industry conferences around the United States since the mid-2000s.

Mr. Duerr holds a Bachelor of Engineering degree from Pratt Institute and a Master of Science in Civil Engineering degree from the University of Houston. Over the course of his career, Mr. Duerr has held professional engineer licenses in twenty-one states. He is a member of the American Society of Civil Engineers, the American Society of Mechanical Engineers and the Society of Automotive Engineers, and is listed in a number of biographical references, including *Who's Who in America* and *Who's Who in Science and Engineering*. In 2012, Mr. Duerr was awarded the ASME Safety Codes and Standards Medal for his work with ASME in the development of lifting equipment design and safety standards.

Preface to the Second Edition

The first edition of this handbook was met with enthusiasm among mobile crane owners and users when published in January of 2015. Along with this acceptance came a continuous stream of suggestions for additional topics that could be addressed and the occasional request for clarification of the existing material. Thus, less than a year after the release of the first edition of the *Mobile Crane Support Handbook*, work began on this second edition.

The basic scope of the book is unchanged. The main focus is on the analysis of crane mats, augmented by presentations of methods used to calculate the support loads imposed by mobile cranes, a general discussion of the determination of soil bearing capacity, and techniques used to analyze subsurface structures. Each of these topics has been expanded upon in this edition of the handbook.

The discussion of crane support loads in Chapter 1 has been expanded to provide examples of support load patterns from crawler and outrigger-supported cranes. Understanding these patterns can give the lift planner a better feel for how changes in crane operations can alter the loads to the supports. The crawler track bearing pressure calculation discussion now provides details of a second calculation method, one that is used by some of the commercial track bearing pressure calculation software products currently available. The discussion of the blocking of crawler tracks has been greatly expanded, including the addition of two examples to illustrate the calculation methods used to determine the track bearing pressures when either one or both ends of the tracks are blocked.

In addition to buried pipes, Chapter 2 now briefly addresses other underground structures, including utility vaults and duct banks, and structures independent of the soil, such as buildings and bridges, on which a crane may be positioned. Additional example problems have been added to illustrate and clarify the calculation methods discussed in the text. The short discussion of methods that can be used to improve ground bearing capacity has been moved from Chapter 5 to Chapter 2 to consolidate the material on bearing capacity. This discussion has also been expanded to address additional methods of improvement. A new section on the bearing capacity of frozen soil has also been added. The conversions to SI units of soil properties that were sourced in U.S. customary units (USCU) follow the rounding conventions given in ASTM/IEEE SI 10-2016 *American National Standard for Metric Practice*. In recognition of the growth of "non-soil" material in this chapter, the chapter title has been revised.

The first section of Chapter 3 in which the results of the timber mat dimension study is reported is largely unchanged from the first edition. There are no new mat measurements to present, although the analyses of the related mat dimensions have been revised to follow the USCU-to-SI conversions and rounding conventions given in SI 10-2016. The mat species table has been expanded due to the inclusion of a few more domestic vendors in the timber species identification study, but the conclusions drawn from this study with respect to the types of wood used for mat construction have not changed. The discussion of allowable stresses has been updated as necessary based on new editions of the *National Design Specification*® *for Wood Construction* and ASTM D2555 *Standard Practice for Establishing Clear Wood Strength Values*, resulting in some very minor changes in the allowable stresses for mora timbers. The later sections of Chapter 3 have been expanded to provide properties and allowable stresses for additional tropical hardwood species that are used for crane mat construction, particularly by mat vendors located outside of the U.S.

The material in Chapter 4 on alternate mat arrangements has been expanded to provide more general solutions to some of the more common arrangements used in practice and to provide additional examples to illustrate the calculation methods, including the use of steel plates over timber mats and steel plates bearing directly on the ground. Also discussed are some mat layout configurations that are often suggested but that do not provide efficient results.

The primary purpose of Chapter 5 remains the same, serving as a summary of how to apply the material developed in Chapters 1 through 4. The example mat design problem has been expanded to show analyses of two different arrangements with two layers of mats. A new section has been added in which some common rules of thumb are discussed, with an emphasis on those that are not valid. The chapter also includes a study of outrigger loads for small and medium sized [150 tons (136 tonnes) and less] outrigger-supported cranes and a brief discussion about inspection of crane mats.

Each chapter now concludes with a section titled Application Summary. These short sections clarify the specific material, equations, and methods that should be used in the design of mobile crane supports. The purpose of these sections is to differentiate between material in each chapter that serves as background or derivation from material that is appropriate for practical use by the lift planner/engineer. The addition of these sections has been driven by users of the first edition of the book who tried to use background material in crane support design, with the results often being somewhat confusing or conflicting.

A new Appendix 3 "Rounding Conventions" has been added at the back of the book. This appendix provides the basis for the rounding of values for conversions from U.S. customary units to SI units and rounding in the development of timber design properties. The standards upon which the information in this appendix is based are cited for the benefit of the user in need of additional detail.

One effort that was constant throughout the book during the writing of this second edition was the refining of the symbols used in the equations to further improve agreement with the notation used in previously published work and industry standards and, where practical, to reduce the number of symbols that have more than one meaning. Completely eliminating these double uses of symbols while still maintaining consistency with other publications was not possible, so a new Appendix 4 "Notation" has been added. This appendix lists all of the symbols used in the equations and figures throughout the book and should assist the reader in understanding the correct usage.

As in the first edition, proprietary products are discussed in some sections as a part of the presentation of this material. The use of such products as examples is for illustrative purposes only and should not be interpreted as an endorsement or recommendation.

Last, the closing paragraph of the preface to the first edition must be repeated here due to its fundamental importance: The design methods presented and developed in this handbook consider the more commonly encountered types of cranes, types of crane mats, and job site conditions. The lift planner/engineer will, of course, come across unique situations that differ from those used here as examples. Thus, this material must not be viewed as a "cookbook" of mobile crane support design. Rather, it is necessary to understand the underlying *principles* to facilitate adaptation to any crane support application. Only then can both safety and practicality be maximized.

David Duerr, P.E.

Houston, Texas
January 16, 2019

Preface to the First Edition

Crane mats are used to distribute the high concentrated loads from a mobile crane over a relatively large area so as to load the soil or other underlying surface at tolerable bearing pressures, as has been common construction industry practice for many decades. The procedures used to evaluate the strength of the mats and to determine the ground bearing pressure under the mats are typically very simple and are based on the assumption of a uniform distribution of bearing pressure. Although crane mats are most commonly made of heavy timbers, fabricated steel mats or mats made of high-strength plastics are occasionally used under very large cranes or when soil conditions are particularly poor.

As cranes have grown in size, the loads imposed on their supports have likewise grown. This has resulted in the need to reevaluate the approaches used to design crane supports. The rules of thumb and simplified methods that were practical with smaller cranes often do not offer the safest and most efficient designs with the larger cranes common today.

The purpose of this handbook is to present mobile crane support design procedures that utilize practical soil and mat properties and that account for the interaction between mats and soil, thus providing a more reliable and realistic design. The need in the construction industry is to have a design approach that can be performed using the crane loading and soil information that is commonly available. More sophisticated analysis procedures, while presumably more accurate than the approach presented here, are often not practical due to the lack of the needed data.

There are many books available today about lifting with mobile cranes and most of these address crane support to varying degrees. What is unique about this handbook is the presentation of original research and rigorous engineering derivations to provide the technical basis for the crane support design methods that are developed. This allows not only laying out *how* to design mobile crane support, but also clearly defining *why* certain equations, allowable stresses, etc. are suggested, thus giving the lift planner/engineer a much improved understanding of the techniques being used.

Methods of calculating the support reactions from a mobile crane are discussed in Chapter 1. This material covers both manual calculations and the use of the computer software products available from the various crane manufacturers and independent developers. Chapter 2 presents a general overview of geotechnical exploration and allowable ground bearing pressures, including considerations of subsurface pipes and structures. Chapter 3 introduces two studies conducted by the author by which the expected strength of the types of timbers used in crane mats has been established. Also addressed briefly is the design and use of crane mats made of materials other than wood. Chapter 4 presents the derivation of a technically sound method of calculating the effective bearing area of a crane mat. This chapter builds upon a well received paper by the author titled "Effective Bearing Length of Crane Mats," which was originally published in 2010 in conjunction with Maximum Capacity Media's Crane & Rigging Conference in Houston. Last, Chapter 5 pulls everything together and outlines the basic crane support design process.

Example problems are used where appropriate to demonstrate and clarify the principles discussed and to illustrate the use of some of the equations presented. The problems are solved primarily using U.S. customary units (USCU), as most of the research upon which this work is based was conducted in the United States and was reported in USCU. In some cases, problem results are summarized in international system (SI) equivalent values where useful for clarity. However, the equations presented throughout this book are dimensionally independent unless otherwise indicated and can be used with either USCU or SI units. It must also be noted that the calculations in the example problems were solved using spreadsheet templates, so the results shown often differ slightly from what one would obtain using pencil and paper due to the number of significant figures carried through by the spreadsheet equations.

In some sections, proprietary products are discussed as a part of the presentation of this material. The use of such products as examples is for illustrative purposes only and should not be interpreted as an endorsement or recommendation.

The design methods presented and developed in this handbook consider the more commonly encountered types of cranes, types of crane mats, and job site conditions. The lift planner/engineer will, of course, come across unique situations that differ from those used here as examples. Thus, this material must not be viewed as a "cookbook" of mobile crane support design. Rather, it is necessary to understand the underlying *principles* to facilitate adaptation to any crane support application. Only then can both safety and practicality be maximized.

David Duerr, P.E.

Houston, Texas
January 5, 2015

1 Crane Loads

The first step in the engineering of the support of a mobile crane is the determination of the reactions imposed by the crane on its supporting surface. These reactions are either concentrated loads from outriggers or bearing pressures under crawler tracks. Methods of calculating these crane support reactions are developed in this first chapter.

While these calculation methods are useful to understand, their use in day to day lift planning is becoming increasingly rare. Most of the major crane manufacturers now produce computer applications, either locally installed or web-based, that provide quick and reliable calculation of support reactions. There are also applications on the market from independent developers that have support reaction calculation routines. Even in the absence of a need to perform these calculations "the old fashioned way," there is still value in understanding the principles and the mathematics behind the computer applications.

1.1 LOADS FROM THE CRANE

Calculation of the support reactions of a mobile crane is accomplished by first reducing the weights of the crane and its lifted load to a total load and two horizontal eccentricities with respect to the center of the support footprint. Simple rules of statics require that the total weight of the crane plus load acting downward must equal the sum of the reactions acting upward and that the center of gravity (CG) of the crane plus load must be in line with the centroid of the support reactions.

Calculation of the weight and center of gravity location of the crane and load is most conveniently done by separating the crane into its two major components. These are the rotating superstructure and the lower structure, whether the carrier of a truck crane or the tracks and carbody of a crawler crane. This approach facilitates calculation of the support reactions as the superstructure swings during a lifting operation.

1.1.1 Reactions from the Superstructure

In this discussion, the term "superstructure" is used to refer to the complete portion of the crane that rotates. This includes the main unit that contains the operating

1

machinery, the counterweights, the boom, the jib (if present), the complete boom hoist system (A-frame, pendants, etc.), and any other components that are part of the crane. The basic superstructure is also referred to as the house, the machinery house, the upper, or the upperworks.

The first calculations combine these various weights and center of gravity positions into a single weight and center of gravity position. The product of this set of calculations is a weight W_s, equal to the sum of the weights of the rotating crane components, and a center of gravity position L_s, measured forward horizontally from the center of rotation in a direction parallel to the boom.

The component centers of gravity and center of gravity locating dimensions of a basic mobile crane superstructure are illustrated in Fig. 1.1. The crane superstructure shown here is that of a common mobile crane with a lattice boom. Other crane models may have auxiliary counterweights, more complex boom

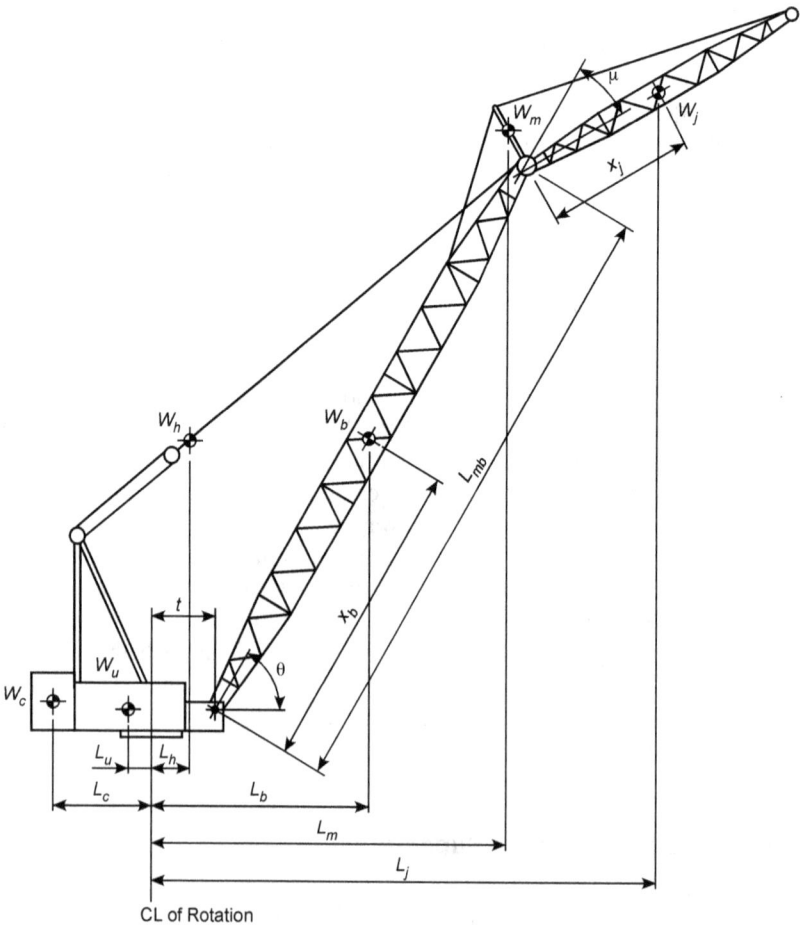

Figure 1.1 Component Weights and Dimensions of a Mobile Crane Superstructure

attachments, or other appurtenances. However, the basic procedure shown here still applies regardless of the complexity of the superstructure's configuration. That is, one must determine the weight and center of gravity location for each of the various components that make up the superstructure.

Once the required information is in hand, the required weight W_s and center of gravity position L_s are calculated using Eqs. 1.1 and 1.2, respectively.

$$W_s = W_u + W_c + W_b + W_h + W_j + W_m \tag{1.1}$$

$$L_s = \left(W_u L_u + W_c L_c + W_b L_b + W_h L_h + W_j L_j + W_m L_m \right) / W_s \tag{1.2}$$

where

W_u	=	weight of the machinery house;
L_u	=	longitudinal center of gravity of the machinery house;
W_c	=	weight of the counterweight;
L_c	=	longitudinal center of gravity of the counterweight;
W_b	=	weight of the boom;
L_b	=	longitudinal center of gravity of the boom;
W_h	=	weight of the boom hoist components;
L_h	=	longitudinal center of gravity of the boom hoist components;
W_j	=	weight of the jib;
L_j	=	longitudinal center of gravity of the jib;
W_m	=	weight of the jib mast and backstays; and,
L_m	=	longitudinal center of gravity of the jib mast and backstays.

The values of the longitudinal center of gravity dimensions are positive when the center of gravity is forward of the center of rotation and negative when the center of gravity is behind the center of rotation. As illustrated in Fig. 1.1, the values of L_c and L_u are negative and the values of L_b, L_h, L_j, and L_m are positive. Great care must be taken when assembling these numerical values to assure that the correct signs are used.

The longitudinal center of gravity lengths for the boom and jib are functions of the boom angle θ and jib offset angle μ. These two values are calculated using Eqs. 1.3 and 1.4, respectively.

$$L_b = t + x_b \cos\theta \tag{1.3}$$

$$L_j = t + L_{mb} \cos\theta + x_j \cos(\theta - \mu) \tag{1.4}$$

where

t	=	horizontal distance from the CL of rotation to the boom foot pin;
x_b	=	boom CG location measured along the length of the boom;
L_{mb}	=	main boom length; and,
x_j	=	jib CG location measured along the length of the jib.

As with the boom and jib center of gravity positions, the center of gravity positions for the boom hoist components L_h and for the jib mast and backstays L_m will also change as the crane is boomed up or down. However, the calculation of these values, particularly L_h, may not be as straightforward due to differences in the component arrangement on various crane models. These values must be determined based on the actual configuration of the crane at hand.

When considered in plan view, the centers of gravity of the components of the superstructure reasonably may be considered to be centered transversely. The counterweight, boom, and jib will be exactly centered, but the weight of the machinery house may be offset to one side or the other. Relative to the overall weight and transverse position of the center of gravity of the crane and the lifted load, this offset typically is not significant. Therefore, only the one center of gravity position L_s need be calculated.

1.1.2   Reactions from the Lower

Calculation of the weight W_l and the center of gravity position L_l of the crane's lower is often simpler than the calculations required for the superstructure. For some crane models, the total weight of the undecked lower is given in the manufacturer's specifications. For cranes with removable components, such as a front bumper (for truck cranes) or carbody (for crawler cranes) counterweight, calculations similar in format to those performed for the superstructure will be required.

The two values required for the support load calculations are shown in Fig. 1.2 for a truck crane and in Fig. 1.3 for a crawler crane. The center of gravity dimension L_l is taken as positive when the center of gravity is toward the front

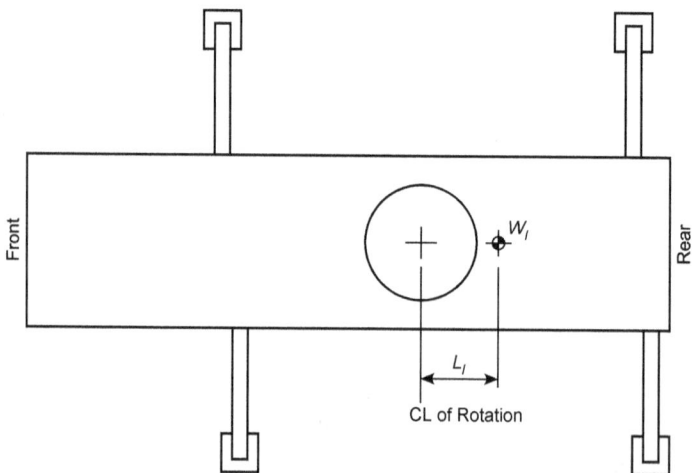

Figure 1.2 Weight and Center of Gravity Location of a Truck Crane Carrier

for a crawler crane and when it is toward the rear for a truck crane. The center of gravity positions illustrated in Figs. 1.2 and 1.3 are both shown such that L_l is positive. Also as with the superstructure, the center of gravity is assumed to be centered transversely; any offset of the center of gravity of the lower to one side or the other will be insignificant relative to the total weight and center of gravity position during a lift.

1.1.3   The Total Lifted Load

The total lifted load to be used in the calculation of a crane's support reactions is equal to the total weight of everything suspended from the boom tip and, if applicable, the jib tip. This includes the weights of the lifted payload, the below-the-hook rigging, the load block, the headache ball, and all of the crane's reeving.

Some cranes, such as many models from Manitowoc, consider all weights below the boom tip and jib tip to be part of the lifted load. Other cranes, such as some Link-Belt models, account for the weight of the reeving in the load charts; thus, only excess reeving beyond that required for the rated load shown on the chart is considered to be part of the lifted load. This difference does not come into play when calculating a crane's support reactions. Here, the full weight of the reeving must be included in the weight and center of gravity calculations

The following notation will be used in subsequent calculations for the boom tip and jib tip loads.

W_{bt} =   total load at the boom tip;
L_{bt} =   horizontal distance from the center of rotation to the boom tip;

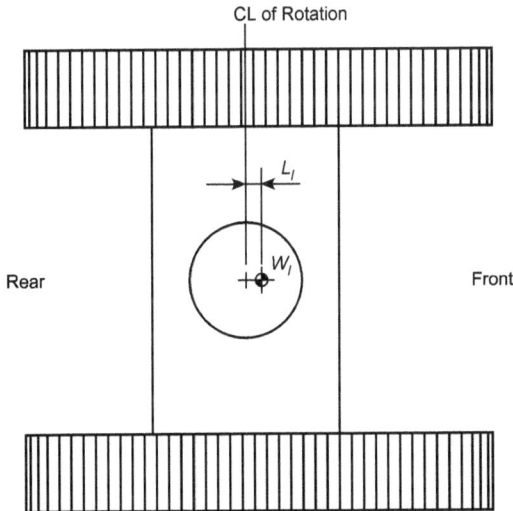

Figure 1.3 Weight and Center of Gravity Location of a Crawler Crane Lower

W_{jt} =   total load at the jib tip; and,
L_{jt} =   horizontal distance from the center of rotation to the jib tip.

Obviously, the horizontal dimension from the center of rotation to the lifted load, which is L_{bt} when lifting from the boom or L_{jt} when lifting from the jib, is the operating radius for the lift.

1.1.4   Where the Challenge Lies

We can see at this point that the calculation of the weights and the center of gravity locations of a crane's superstructure and lower are quite simple. The math is not at all complex and the equations lend themselves to being set up in a spreadsheet or mathematics program, which will further speed the work and reduce the likelihood of computational error. There remains one significant challenge, however.

The value of the results of these calculations is largely dependent upon the accuracy with which the weights and center of gravity positions of the individual components of the crane are known and getting accurate values for the weight and center of gravity position for each major component of a crane may be difficult. The specifications for larger cranes will normally include a list of weights of each component. This information is needed for planning the transportation activities to get the crane to a job site and for lift planning for the assembly and disassembly of the crane. The center of gravity locations for uniform items, such as boom insert sections, can usually be estimated reasonably closely, but determining the center of gravity position for more complex assemblies, such as the machinery house, is more difficult. Determining the center of gravity position for a telescopic hydraulic boom at its various lengths of extension can be very difficult.

Although sometimes difficult, the information needed to perform these calculations can be obtained. The crane's transport documentation should be the first place to look and a specific request to the manufacturer can usually provide any additional numbers needed.

1.2   CRANE SUPPORT LOADS

The calculations developed in Section 1.1 give us the weight and center of gravity position for the crane's superstructure (including the boom and jib), for the lower (carrier or carbody and crawlers), and for the loads suspended from the boom tip and from the jib tip. We will now combine these four loads and center of gravity positions into one composite weight and center of gravity position for the crane with its lifted loads. These numbers then can be resolved into outrigger loads or crawler track bearing pressures.

Two more variables must be introduced for the combining of the three sets of weight values. The first is the angle of rotation (swing angle) of the superstructure, which we will identify as α. A swing angle of zero degrees is often taken as straight over the front for crawler cranes and straight over the rear for outrigger-supported cranes, which is the convention used in this book. As used in these calculations, the value of the angle increases as the superstructure swings counterclockwise. The second variable is the longitudinal offset of the centerline of the outrigger group or crawler track footprint to the center of rotation x_o. This value is positive when the centerline of the outrigger group is to the rear of the center of rotation, as shown in Fig. 1.4, and positive when the centerline of the crawler track footprint is forward of the center of rotation, as shown in Fig. 1.5.

With the established notation and the swing angle defined, we can write Eqs. 1.5 through 1.7 for the calculation of the total vertical load V and the longitudinal and transverse centers of gravity, LCG and TCG respectively, relative to the centroid of the outriggers or crawler footprint.

$$V = W_l + W_s + W_{bt} + W_{jt} \tag{1.5}$$

$$LCG = \frac{W_l L_l + \left(W_s L_s + W_{bt} L_{bt} + W_{jt} L_{jt}\right)\cos\alpha}{V} - x_o \tag{1.6}$$

$$TCG = \frac{\left(W_s L_s + W_{bt} L_{bt} + W_{jt} L_{jt}\right)\sin\alpha}{V} \tag{1.7}$$

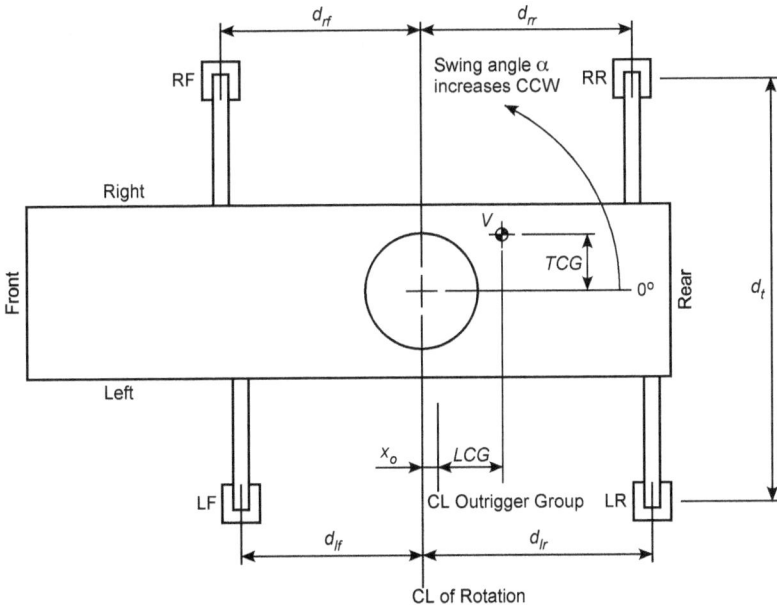

Figure 1.4 Dimension Notation for Outrigger Load Calculations

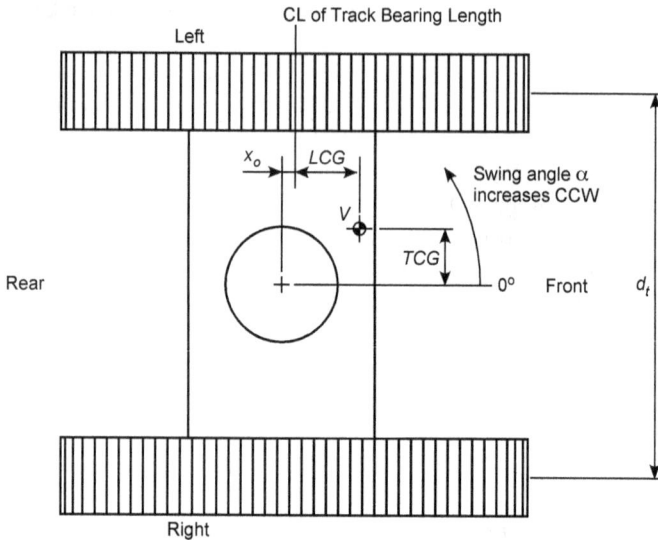

Figure 1.5 Dimension Notation for Crawler Track Load Distribution

When applying Eqs. 1.6 and 1.7 to an outrigger-supported crane, *LCG* is positive to the rear and *TCG* is positive to the right, as illustrated in Fig. 1.4. When applying Eqs. 1.6 and 1.7 to a crawler crane, *LCG* is positive to the front and *TCG* is positive to the left, as illustrated in Fig. 1.5. In both cases, "left" and "right" are reckoned from the position facing the front of the crane. And of course, the application of Eq. 1.5 is consistent for all types of cranes.

1.2.1   Outrigger Loads

The calculation of outrigger loads by necessity follows a relatively simplistic approach. There are two methods that are in common use today, referred to here as the Rigid Body method and the Geometric Distribution method. Both methods are described in this section.

Rigid Body. This calculation method treats the crane as a rigid body supported at four points and loaded by a vertical load and two orthogonal moments, where the moments are equal to the vertical load *V* multiplied by *LCG* or *TCG*. The solution is often further simplified by ignoring the longitudinal offsets of the two pairs of outriggers and using the average distance from the center of the front outrigger pair to the center of the rear outrigger pair d_l (Eq. 1.8).

$$d_l = \frac{d_{rr} + d_{lr}}{2} + \frac{d_{rf} + d_{lf}}{2} \qquad (1.8)$$

This simplification of the longitudinal spacing of the outriggers is used, for example, in Grove's Compu-Crane web site calculator for some crane models. We can see this in the results shown in Fig. 1.6. The printout in Fig. 1.6*a* shows the outrigger loads for a Grove GMK5175 crane lifting straight over the rear. The two front outrigger loads are equal to one another, as are the rear outrigger loads. This shows that the staggered longitudinal spacing of the outriggers is not accounted for in these calculations. For comparison, Fig. 1.6*b* shows the outrigger loads for a Grove RT890E crane, also lifting straight over the rear. Here, the front outrigger loads differ from one another, as do the rear outrigger loads. This shows that the staggered longitudinal spacing has been accounted for in the calculations of the outrigger loads for this crane model.

As before, we can take the sum of the weights of the crane and its lifted loads as V, where V is given by Eq. 1.5. With these variables now defined, we can write Eqs. 1.9 through 1.12 for the calculation of the four outrigger loads.

GMK5175 w/ Main Boom Only - English Outrigger Pad Loads

Always confirm that the lifted load and configuration are approved in the load chart.

Boom Length: 101.7 ft   [100-100-0-0]
    Counterweight: 99200# Cwt
     Outriggers: 28.1 x 26.9
     Load Radius: 30 ft
     Load Weight: 100000 lbs

Slew Angle: 0° = Directly Over Rear
    33700 lbs Fwd-Lt
    33700 lbs Fwd-Rt
    130200 lbs Aft-Lt
    130200 lbs Aft-Rt

(a) Outrigger Loads for a Grove GMK5175 Crane

RT890E Outrigger Pad Loads
Always confirm that the lifted load and configuration are approved in the load chart.
    Main Boom: 37.3 ft - 141.7 ft Main Boom, Mode 'B' @ 102.81 ft. length [100-100-25-25]
Superstructure: Std superstructure with main hoist + cable
Counterweight: Removable Cwt with cyl. and IPO cwt.
Lift Cylinder: Std. Lift Cyl. Data
    Carrier: Standard Carrier
  Outriggers: Fully Extended Outriggers
  Hook Load: 33000 lbs. including rigging Radius: 35 ft.
Slew Angle: 0 deg. = Directly Over Rear
  12182.54 lbs. Fwd-Lt
  10641.48 lbs. Fwd-Rt
  60479.17 lbs. Aft-Lt
  61661.82 lbs. Aft-Rt

(b) Outrigger Loads for a Grove RT890E Crane

Figure 1.6 Example Outrigger Load Calculation Results

$$P_{RR} = \frac{V}{4} + \frac{V(LCG)}{2d_l} + \frac{V(TCG)}{2d_t} \qquad (1.9)$$

$$P_{LR} = \frac{V}{4} + \frac{V(LCG)}{2d_l} - \frac{V(TCG)}{2d_t} \qquad (1.10)$$

$$P_{RF} = \frac{V}{4} - \frac{V(LCG)}{2d_l} + \frac{V(TCG)}{2d_t} \qquad (1.11)$$

$$P_{LF} = \frac{V}{4} - \frac{V(LCG)}{2d_l} - \frac{V(TCG)}{2d_t} \qquad (1.12)$$

This calculation method, in some cases, will show a negative value at an outrigger. This, of course, is not possible; outriggers are not tension anchors. This result indicates that one outrigger will lift off the supporting surface. In this case, the support reactions must be re-calculated based on a three-point support, ignoring the outrigger that showed the negative value.

Geometric Distribution. This method simply divides the total load front to rear and left to right based on the position of the center of gravity of the total crane with load relative to the four outriggers. Using the same notation as was used for the Rigid Body method, we can write Eqs. 1.13 through 1.16.

$$P_{RR} = V\left(\frac{d_t/2 + TCG}{d_t}\right)\left(\frac{d_{rf} + x_o + LCG}{d_{rf} + d_{rr}}\right) \qquad (1.13)$$

$$P_{LR} = V\left(\frac{d_t/2 - TCG}{d_t}\right)\left(\frac{d_{lf} + x_o + LCG}{d_{lf} + d_{lr}}\right) \qquad (1.14)$$

$$P_{RF} = V\left(\frac{d_t/2 + TCG}{d_t}\right)\left(\frac{d_{rr} - x_o - LCG}{d_{rf} + d_{rr}}\right) \qquad (1.15)$$

$$P_{LF} = V\left(\frac{d_t/2 - TCG}{d_t}\right)\left(\frac{d_{lr} - x_o - LCG}{d_{lf} + d_{lr}}\right) \qquad (1.16)$$

Comparison of this set of equations to the first set will reveal a significant difference. Unlike the Rigid Body method, outrigger load calculation by the Geometric Distribution method will never produce a zero or a negative outrigger load. That is, all four outriggers will always carry some load, regardless of the crane's swing angle or loading.

Comparison of the Calculation Methods. An evaluation of these two calculation methods can be made by using crane manufacturers' software to compute the outrigger loads for a particular crane and then comparing those outrigger loads to the loads calculated using both of the methods developed here. The approach used in this evaluation is as follows:

1. Input a crane configuration and the particulars of a lift into the crane manufacturer's software and compute the outrigger loads.
2. Use these outrigger loads to compute the gross weight (V) and center of gravity location (LCG and TCG) of the crane with its lifted load.
3. Use these values in Eqs. 1.9 through 1.12 to calculate the outrigger loads by the Rigid Body method and in Eqs. 1.13 through 1.16 to calculate the outrigger loads by the Geometric Distribution method.
4. Compare the three sets of outrigger loads, particularly noting the maximum and minimum outrigger loads produced by each method.

In the calculations developed in this book, zero-degree swing is over the rear and the swing angle increases in a counterclockwise direction. Note that these conventions are not consistent among the crane manufacturers' software. For example, zero degrees swing is over the front in the Kobelco software and the swing angle increases in the clockwise direction in the Grove software. As always, care must be taken to be sure that the conventions are clearly understood when using these software products.

This comparison process has been performed by the author for dozens of lift configurations using a range of crane models from three different manufacturers (Grove, Kobelco, and Link-Belt). Interestingly, neither of the calculation methods described here agreed exactly with the manufacturers' software and neither showed consistent approximate agreement. Some of the cases checked produced loads at only three outriggers (that is, one outrigger showed a zero load). For these cases, the Rigid Body method produced closer results, since the Geometric Distribution method never shows a zero load. In some cases, the manufacturer's software showed a zero outrigger load, but the Rigid Body calculation showed a small load at the same location. When the Rigid Body calculation also showed a zero load, the values of the other three outrigger loads agreed exactly with those from the manufacturer's software. Otherwise, the agreement with the outrigger loads calculated with the manufacturers' software was split roughly equally between the two methods.

This lack of exact agreement is not necessarily a cause for concern. We don't know the specifics of the calculation routines used by the manufacturers' software. For example, we don't know what kind of consideration is given to flexure of the carrier and outrigger beams. (Both of the calculation methods described here are based on the crane's dimensions only; structural deformation of the crane is not taken into account.)

The differences in the outrigger loads calculated by either method described here compared to the loads from the manufacturers' software are relatively small, typically less than 10% and never more than 15%. More importantly, the differences where the maximum outrigger load calculated using the methods developed in this chapter is less than the maximum outrigger load from the manufacturer's software were found to be about 5% or less and in these cases, the Geometric Distribution method showed somewhat smaller differences than the Rigid Body method.

Given this examination of calculation results, the Rigid Body method appears to be preferable, but either method can be used without risk of great inaccuracy. This statement notwithstanding, however, the use of outrigger loads calculated using the crane manufacturer's software is still the first choice.

Front Outriggers. Calculation of outrigger loads, regardless of the method used, becomes more complex when the crane is equipped with a front outrigger. One approach to this problem is to assume a load that will act at the front outrigger and incorporate that value into the calculation of the weight and center of gravity location of the crane. In essence, the assumed front outrigger load is treated as a negative (upward) force acting on the crane.

Some crane manufacturers' outrigger load calculation software products use this approach. The front outrigger is assigned a fixed load, regardless of front bumper counterweight size, swing angle of the superstructure, or load being lifted. Guidance on the appropriate front outrigger reaction to use in lift planning must be sought from the crane's manufacturer.

1.2.2   Crawler Track Bearing Pressures

The calculation of the track bearing pressures imposed by a crawler crane on its supporting surface requires the introduction of a number of dimensions. The distribution of the total vertical load V between the two tracks is simply based on the transverse center of gravity TCG and the center-to-center spacing d_t of the crawler tracks, illustrated in Fig. 1.5. The distribution of the bearing pressure along the length of each track is based on the longitudinal center of gravity LCG (Fig. 1.5), the length of bearing of the tracks and the width of bearing of the tracks, the dimensions of which require additional explanation. Before proceeding to the calculations, we must first develop an understanding of these length and width of bearing dimensions.

Of particular importance with many crawler crane models is the distinction between the dimensions of the track bearing areas on a soft surface vs. those on a hard surface. When bearing on a soft surface, which generally means bare soil, the material will compress locally and allow the tracks to bear over their full length [approximately center-to-center of the sprockets (also called tumblers)] and over the full width of the treads (also called track pads or track shoes). This behavior along the length of the tracks is illustrated in Fig. 1.7a. This local compression will not occur on a hard surface, such as crane mats, thus limiting the length of the track bearing area to the length along the track rollers (Fig. 1.7b). The lack of bearing beyond the track rollers can be seen in the photograph of the track of a Liebherr LR1300 crane on timber mats in Fig. 1.8.

Looking at the width of the track, when bearing on a soft surface, this local compression will allow the treads to bear across their full width (Fig. 1.9a). On a hard surface, bearing will be limited to the flat center section of the treads (Fig. 1.9b).

For reference, not all crawler cranes have the sprockets raised at each end, as is shown in Figs. 1.7a and 1.7b. Also, the tread shape used in Fig. 1.9 is based on the dimensions of the standard treads on the Manitowoc 2250 crane. Some treads, including those on some other Manitowoc crane models, are flat across their full width. Thus, not every crawler crane presents different track bearing surface areas for soft vs. hard surfaces. However, for cranes that do have these different dimensions, the correct dimensions must be used in the track bearing pressure calculations.

With this explanation of track bearing areas in place, we can now define the actual dimensions to be used in the bearing pressure calculations. The track length dimensions are illustrated in Fig. 1.10 and the tread width dimensions are illustrated in Fig. 1.11. All of the dimensions shown in Figs. 1.5, 1.10 and 1.11,

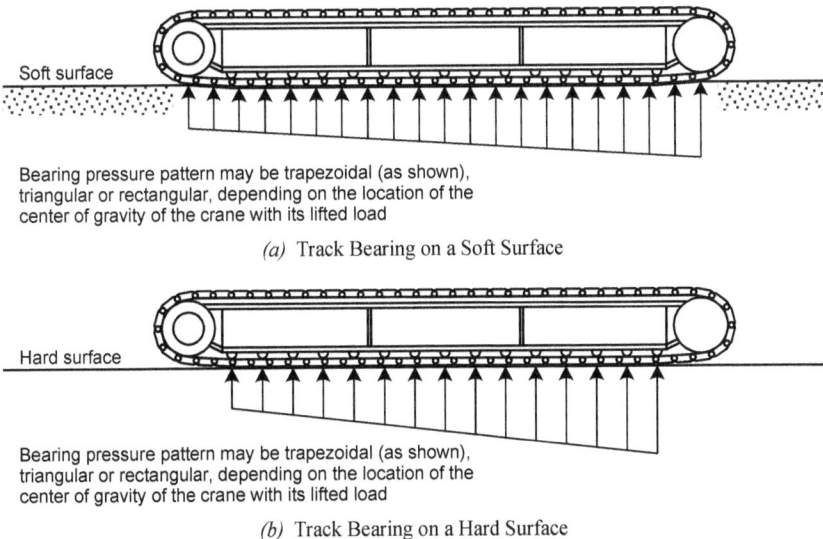

Bearing pressure pattern may be trapezoidal (as shown), triangular or rectangular, depending on the location of the center of gravity of the crane with its lifted load

(a) Track Bearing on a Soft Surface

Bearing pressure pattern may be trapezoidal (as shown), triangular or rectangular, depending on the location of the center of gravity of the crane with its lifted load

(b) Track Bearing on a Hard Surface

Figure 1.7 Crawler Track Bearing Lengths

Figure 1.8 Crawler Track on Timber Mats *(Chevron Products Company)*

except the *LCG* and *TCG* values in Fig. 1.5, are found in the crane manufacturer's specifications for the particular model. *LCG* and *TCG* are calculated for the specific crane configuration and lift using Eqs. 1.6 and 1.7, respectively.

The total vertical load *V* (Eq. 1.5) is distributed to the two crawler tracks as a function of the transverse center of gravity position *TCG*. The left and right crawler track loads, noted as P_l and P_r respectively, are calculated with Eqs. 1.17 and 1.18.

$$P_l = V \frac{d_t/2 + TCG}{d_t} \tag{1.17}$$

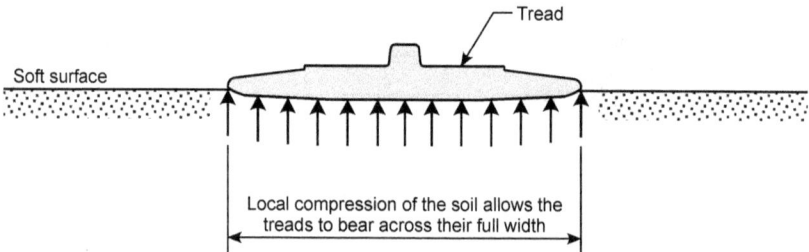

(a) Track Bearing on a Soft Surface

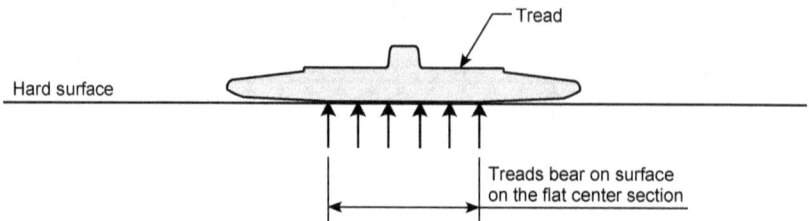

(b) Track Bearing on a Hard Surface

Figure 1.9 Crawler Track Bearing Widths

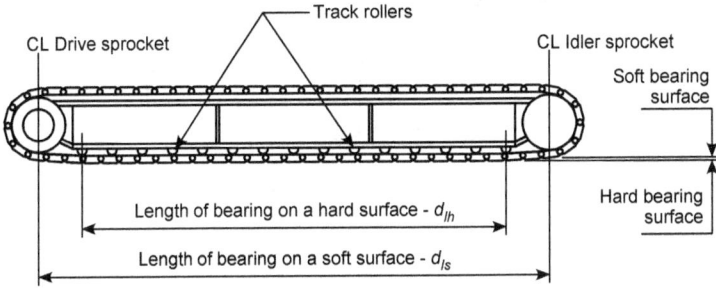

Figure 1.10 Crawler Track Bearing Length Dimensions

$$P_r = V \frac{d_t/2 - TCG}{d_t} \tag{1.18}$$

The shape of the bearing pressure envelope over the length of the track may be trapezoidal, triangular, or rectangular. The difference depends on how far off center longitudinally the center of gravity of the load acting on the track is located. Further, the difference depends on the method used to calculate the track bearing pressure envelope. There are two methods that are currently used in practice. The simpler of the two methods assumes that the value of LCG calculated for the crane as a whole applies equally to both tracks. In this case, the length of bearing will be identical for both tracks. The second, more complex, method assumes that the rate of change of the bearing pressure is the same for both tracks. The result of this method is that the value of LCG is often different for each track and the lengths of bearing for the two tracks may be different when the bearing pressure envelope is triangular. The necessary equations for both methods are developed in the following sections.

The notation used in these calculations is illustrated in Fig. 1.12 for the trapezoidal bearing pressure envelope and in Fig. 1.13 for the triangular bearing pressure envelope. The notation of Fig. 1.12 also applies to the rectangular bearing pressure envelope, for which the eccentricity e is equal to zero. Last, although the figures show hard surface bearing (upper bound $L = d_{lh}$ and $C = w_h$), the same

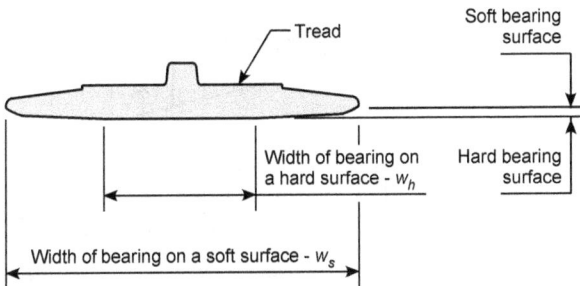

Figure 1.11 Crawler Track Bearing Width Dimensions

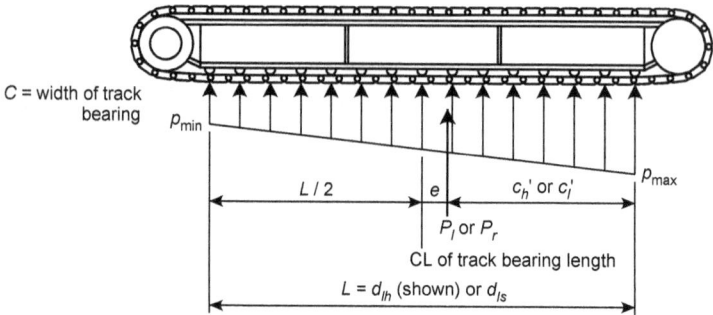

Figure 1.12 Crawler Track Bearing Notation - Trapezoidal Pressure Diagram

notation applies to an analysis for a soft surface (upper bound $L = d_{ls}$ and $C = w_s$) as well.

In the following calculations, LCG is taken as the absolute value of the actual LCG. If the actual value of LCG is positive, then the maximum bearing pressures occur at the front ends of the tracks. If the actual value of LCG is negative, then the maximum bearing pressures occur at the rear ends of the tracks.

Equal LCG for Both Tracks. Using this method, the shape (rectangular, trapezoidal, or triangular), the length of the bearing pressure envelope, and the eccentricity e of the track load will be the same for both tracks and e of the bearing pressure envelope will be equal to LCG of the crane overall. If e is zero, then the shape of the bearing pressure envelope is rectangular and the track bearing pressure is uniform along the track bearing length, equal to P/CL. If e is equal to or greater than one-sixth of the upper bound track bearing length, then the bearing pressure envelope is triangular, the actual length of bearing L is given by Eq. 1.19, the maximum track bearing pressure p_{max} is given by Eq. 1.20, and the minimum track bearing pressure p_{min} is equal to zero. Otherwise, the bearing pressure envelope is trapezoidal, the actual length of bearing L is equal to the upper bound track bearing

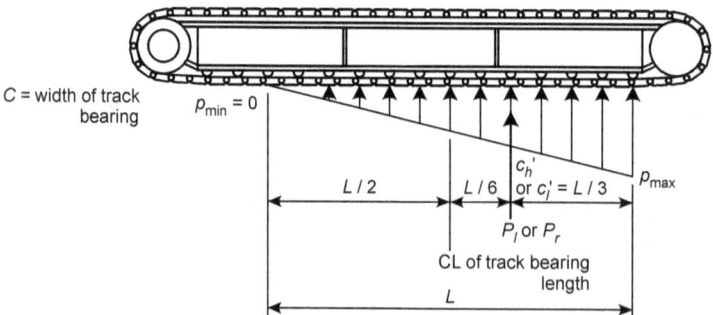

Figure 1.13 Crawler Track Bearing Notation - Triangular Pressure Diagram

length d_l, and the maximum track bearing pressure p_{max} and the minimum track bearing pressure p_{min} are given by Eq. 1.21.

$$L = 3\left(\frac{d_l}{2} - LCG\right)$$
(1.19)

$$p_{max} = 2\frac{P}{CL}$$
(1.20)

$$p = \frac{P}{CL}\left(1 \pm \frac{6e}{L}\right)$$
(1.21)

where
d_l	=	upper bound bearing length of track;
	=	d_{ls} or d_{lh}, as applicable;
P	=	track load;
	=	P_l or P_r, as applicable;
C	=	width of track bearing;
	=	w_s or w_h, as applicable;
p	=	p_{max} when the operation in the brackets is +;
p	=	p_{min} when the operation in the brackets is -; and,
		all other terms are as previously defined.

This approach to the calculation of track bearing pressures will yield the same length of bearing L for both tracks and the eccentricity e will equal LCG for both tracks, regardless of the swing angle of the superstructure. Some of the crane manufacturers' applications available to calculate crane support reactions are based on this assumption of equal track bearing length. Output from Link-Belt's Ground Bearing Information web site calculator (Fig. 1.14) illustrates the form of

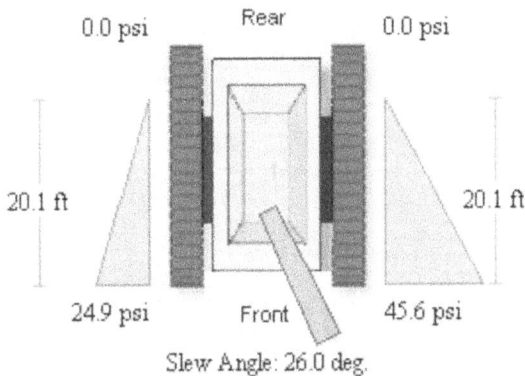

Figure 1.14 Track Bearing Pressures Based on Equal LCG Values

the results obtained using this calculation method. Kobelco's Ground Pressure/
Outrigger Reaction Force web site calculator also uses this method.

Equal Rate of Change for Both Tracks. The loads to each track in this method
are again calculated using Eqs. 1.17 and 1.18. However, the calculation of the
longitudinal pressure distribution along each crawler is performed based on the
assumption of a rate of change of pressure along the length of bearing to be the
same for both tracks. This approach is more complex than the method embodied
in Eqs. 1.19 through 1.21.

There are three possible track bearing pressure envelope configurations that
may occur when the pressures are calculated using this method, illustrated in
Fig. 1.15. Both pressure diagrams may be triangular (Fig. 1.15*a*), both diagrams
may be trapezoidal (Fig. 1.15*b*), or one pressure diagram may be trapezoidal and
the other triangular (Fig. 1.15*c*). All three configurations will be examined in this
section.

Some of the crane manufacturers' applications employ this bearing pressure
calculation method. The graphics in Fig. 1.15 are taken from Manitowoc's Ground
Bearing Pressure Estimator program. Liebherr's LICCON program also uses this
method.

The total crane load *V* and the center of gravity locating dimensions *LCG* and
TCG (Fig. 1.5) are determined as is done in the "equal *LCG*" method. The next
step in this analysis is the determination of the bearing pressure envelope pattern
that is produced by the imposed crane loading. This is accomplished using Eqs.
1.22 through 1.26.

$$c' = \frac{d_l}{2} - LCG \tag{1.22}$$

$$\lambda = \frac{c'}{d_l} \tag{1.23}$$

$$R_T = \text{MAX}\left(\frac{P_l}{V}, \frac{P_r}{V}\right) \tag{1.24}$$

$$\lambda_{tr} = 0.691 R_T^2 - 0.998 R_T + 0.659 \tag{1.25}$$

$$\lambda_{tp} = \frac{R_T}{3} + \frac{1}{6} \tag{1.26}$$

where
c' = longitudinal center of gravity from the heavily loaded end of the
             tracks (this value is always less than or equal to 0.50 d_l);
λ = ratio of c' to the maximum bearing length of the track;

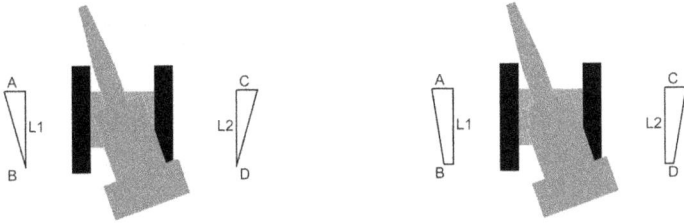

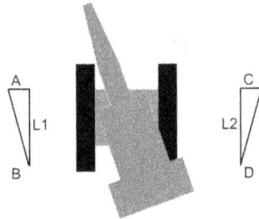

Boom at Critical 19 degree Swing

Center of Rotation to Fulcrum: 136.25 inch
203.1 psi
0.0 psi
265.8 inch
177.9 psi
0.0 psi
232.7 inch

(a)

Boom at Critical 19 degree Swing

Center of Rotation to Fulcrum: 136.25 inch
115.1 psi
48.9 psi
272.5 inch
108.1 psi
41.8 psi
272.5 inch

(b)

Boom at Critical 19 degree Swing

Center of Rotation to Fulcrum: 136.25 inch
179.2 psi
10.4 psi
272.5 inch
159.9 psi
0.0 psi
258.1 inch

(c)

Figure 1.15 Track Bearing Pressures (TBP) Based on Equal Rate of Change of TBP

R_T =   ratio of the higher track load to the total crane load;

λ_{tr} =   the limit of λ that defines the point at which both bearing pressure envelopes are triangular;

λ_{tp} =   the limit of λ that defines the point at which both bearing pressure envelopes are trapezoidal; and,

all other terms are as previously defined.

If λ is less than λ_{tr}, then both bearing pressure envelopes are triangular. If λ is greater than λ_{tp}, then both bearing pressure envelopes are trapezoidal. If λ falls between λ_{tr} and λ_{tp}, then the bearing pressure envelope under the more heavily loaded track will be trapezoidal and the bearing pressure envelope under the opposite track will be triangular.

If the above analysis indicates that both pressure envelopes are triangular, the length of bearing, the maximum bearing pressure, and the rate of change of the bearing pressure for each track are calculated using Eqs. 1.27 through 1.32.

$$c_h' = \frac{Vc'}{T_h + \dfrac{T_l}{\sqrt{T_h/T_l}}} \tag{1.27}$$

$$L_{bh} = 3c_h' \le d_l \tag{1.28}$$

$$c_l' = \frac{c_h'}{\sqrt{T_h/T_l}} \tag{1.29}$$

$$L_{bl} = 3c_l' \le d_l \tag{1.30}$$

$$P_{max} = 2\frac{T}{CL} \tag{1.31}$$

$$R_c = \frac{P_{max}}{L} \tag{1.32}$$

where
c_h'	=	longitudinal center of gravity from the heavily loaded end of the more heavily loaded track;
c_l'	=	longitudinal center of gravity from the heavily loaded end of the more lightly loaded track;
L	=	L_{bh} or L_{bl} as discussed below;
L_{bh}	=	length of bearing of the more heavily loaded track;
L_{bl}	=	length of bearing of the more lightly loaded track;
T	=	T_h or T_l as discussed below;
T_h	=	greater track load; i.e., the greater of P_l or P_r;
T_l	=	lesser track load; i.e., the lesser of P_l or P_r;
R_c	=	rate of change of the track bearing pressure; and, all other terms are as previously defined.

Eqs. 1.31 and 1.32 may be applied to the more heavily loaded track by using $T = T_h$ and $L = L_{bh}$ or the more lightly loaded track by using $T = T_l$ and $L = L_{bl}$.

If the above analysis indicates that both pressure envelopes are trapezoidal, both lengths of bearing are equal to d_l and the bearing pressures and the rate of change of the bearing pressure for each track are calculated using Eqs. 1.33 through 1.40 (Eqs. 1.38 through 1.40 apply to either track by using the appropriate value of T).

$$P_{V\,max} = \frac{V}{2Cd_l}\left[1 + \frac{6(LCG)}{d_l}\right] \tag{1.33}$$

$$p_{V\,min} = \frac{V}{2Cd_l}\left[1-\frac{6(LCG)}{d_l}\right]$$ (1.34)

$$R_{Vc} = \frac{p_{V\,max}-p_{V\,min}}{d_l}$$ (1.35)

$$c_h' = \frac{d_l}{2}-\frac{0.5LCG}{R_T}$$ (1.36)

$$c_l' = \frac{Vc'-T_h c_h'}{T_l}$$ (1.37)

$$p_{max} = \frac{T}{Cd_l}+R_{Vc}\frac{d_l}{2}$$ (1.38)

$$p_{min} = \frac{T}{Cd_l}-R_{Vc}\frac{d_l}{2}$$ (1.39)

$$R_c = \frac{p_{max}-p_{min}}{d_l}$$ (1.40)

where

$p_{V\,max}$ = maximum track bearing pressure, calculated from the total crane weight and longitudinal center of gravity;

$p_{V\,min}$ = minimum track bearing pressure, calculated from the total crane weight and longitudinal center of gravity;

R_{Vc} = rate of change of the track bearing pressure, based on the total crane weight and longitudinal center of gravity; and,

all other terms are as previously defined.

Last, if the above analysis indicates that one pressure envelope is trapezoidal and the other is triangular, the length of bearing L of the track with the triangular bearing pressure envelope, the maximum bearing pressure, and the rate of change of the bearing pressure for each track are calculated using Eqs. 1.41 through 1.50.

$$\text{Assumed } R_c = R_{Vc}\frac{3.036R_T^2-2.362R_T+0.933}{R_T}$$ (1.41)

$$c_h' = \frac{3d_l\left(\frac{T_h}{Cd_l}-R_{Vc}\frac{d_l}{2}\right)+d_l^2 R_{Vc}}{\frac{6T_h}{Cd_l}}$$ (1.42)

$$c_l' = \frac{Vc' - T_h c_h'}{T_l} \qquad (1.43)$$

$$L = 3c_l' \qquad (1.44)$$

$$e_h = \frac{d_l}{2} - c_h' \qquad (1.45)$$

$$\text{Heavy Track } p_{max} = \frac{T_h}{Cd_l}\left(1 + \frac{6e_h}{d_l}\right) \qquad (1.46)$$

$$\text{Heavy Track } p_{min} = \frac{T_h}{Cd_l}\left(1 - \frac{6e_h}{d_l}\right) \qquad (1.47)$$

$$\text{Heavy Track } R_c = \frac{p_{max} - p_{min}}{d_l} \qquad (1.48)$$

$$\text{Light Track } p_{max} = 2\frac{T_l}{CL} \qquad (1.49)$$

$$\text{Light Track } R_c = \frac{p_{max}}{L} \qquad (1.50)$$

Note that this solution is not closed. A bearing pressure rate of change for the crane must be assumed. Eq. 1.41 provides a value that is reasonably close, typically within 2% of the final value of R_c. For most practical applications of these calculations, this degree of accuracy is adequate. If better agreement is required, either the assumed value of R_c can be adjusted manually until the desired level of agreement is reached or a routine can be written in a suitable computer application to perform the necessary iteration.

Bases for the Track Bearing Pressure Calculation Methods. The bases for the two methods of outrigger load calculation are fairly self-explanatory, as they are based simply on the outrigger locating dimensions and the location of the crane's center of gravity. The two methods defined in this section for the calculation of crawler track bearing pressures are somewhat more complex and, thus, call for explanation.

The division of the total crane load between the two crawler tracks is a function of the track spread d_l and the transverse center of gravity dimension TCG. This is a simple statics solution.

The calculation of the track pressures based on the assumption that the locations of the longitudinal center of gravity for both pressure envelopes are equal to the longitudinal center of gravity position for the crane as a whole is straightforward. If the distance from the more heavily loaded end of the track to the centroid of the load on that track is less than one-third of the maximum bearing length of the track (d_{lh} if supported on a hard surface, d_{ls} if supported on a soft surface), then the pressure envelopes will be triangular in shape ($p_{min} = 0$). If the centroid of the load is beyond that point (that is, if the centroid of the load is within the middle third of the bearing length), then the pressure envelopes will be trapezoidal. The solution for the bearing length L and the maximum and minimum bearing pressures p_{max} and p_{min} are given by equations that should be familiar to civil engineers from traditional spread footing design.

The calculation of the track pressures based on the assumption that the rate of change of pressure from p_{max} to p_{min} is the same for both tracks (i.e. R_c from Eq. 1.48 is equal to R_c from Eq. 1.50) is markedly more complex. Some of the equations used in this method were developed by curve fitting.

The form of the bearing pressure envelope combination (triangular-triangular, trapezoidal-trapezoidal, or trapezoidal-triangular) is identified by the ratio of the longitudinal center of gravity as measured from the more heavily loaded ends of the tracks to the maximum bearing length, noted as λ. The values of λ_{tr} and λ_{tp} that separate one region from another were determined by analyzing numerous crane load and track geometry configurations and determining the values of λ that corresponded to the changes in pressure envelope combination. Eqs. 1.25 and 1.26 were developed by fitting curves to these data.

The equations for the triangular-triangular combination (Eqs. 1.27 through 1.32) were derived by analysis of the pair of triangular bearing pressure envelopes. These equations are mathematically exact.

The equations for the trapezoidal-trapezoidal combination (Eqs. 1.33 through 1.40) were derived by analysis of the pair of trapezoidal bearing pressure envelopes. Again, these equations are mathematically exact.

The solution for the trapezoidal-triangular combination (Eqs. 1.41 through 1.50) requires making an initial assumption of the rate of change of the bearing pressure. A formula is presented (Eq. 1.41) by which a value of the rate of change can be estimated. This equation was developed by analyzing numerous crane configurations and fitting a curve to the resulting data. As such (and as noted on page 22), the value obtained from Eq. 1.41 is approximate. The remainder of the equations in this group were derived by analysis of the geometry of the two pressure envelopes and are mathematically exact. Due to the approximation of the initial value of R_c, however, the overall solution is approximate. Greater precision for a particular analysis can be obtained, if necessary, through iteration.

Comparison of the Calculation Methods.   As with the two outrigger load calculation methods, these two track bearing pressure calculation methods yield different results. Unlike the outrigger load calculations, though, the differences are somewhat more complex and may not be obvious to the user.

The location of the centroid of the bearing pressure envelope measured from the point of highest pressure (points A and C in Fig. 1.15) is shown in the equations as c' for the crane as a whole and as c_h' and c_l' for the heavier loaded and lighter loaded tracks, respectively. The value of c_h' is found to be greater in the "equal rate of change" method, compared to the "equal LCG" method. This difference in c_h' results in a flattening of the pressure envelope, which results in a lower maximum bearing pressure in the "equal rate of change" method, even though the total load on the track is the same in both methods. A comparison of track bearing pressures calculated by both methods for a variety of cranes shows that the difference in maximum bearing pressure varies by as much as about 15%, with the value from the "equal LCG" method being consistently greater.

The value of c_l' of the less heavily loaded track is found to be less in the "equal rate of change" method, compared to the "equal LCG" method. This is an expected consequence of the simple fact that the value of c' for the crane as a whole must be the same for both methods. This difference in c_l' results in a steepening of the pressure envelope, which results in a greater maximum bearing pressure in the "equal rate of change" method. Again comparing track bearing pressures calculated by both methods shows that the difference in maximum bearing pressure varies by as much as 50%, this time with the value from the "equal rate of change" method being consistently greater.

Taking these two observations together and with respect to the analyses performed as a part of the development of these calculation methods, we see that the maximum track bearing pressure always occurs under the more heavily loaded track. Thus, even though the maximum bearing pressure under the more lightly loaded track may differ between the two calculation methods by a relatively large percentage, the maximum bearing pressure under the crane will always be that under the more heavily loaded track, where the difference between the two methods is only about 15% at most.

Last, we must note that the results of both calculation methods have been used for mobile crane installation design for decades and both have been shown to produce acceptable results. And again, as with the calculation of outrigger loads, the use of track bearing pressures calculated using the crane manufacturer's software is preferred for crane support design work.

1.2.3   Methods for Reducing Track Bearing Pressures

The track proportions of some crawler cranes result in significantly greater track bearing pressures when the crane is set up on a hard surface, relative to the pressures that occur for a soft bearing surface. Certain methods can be employed in some situations to reduce the track pressures when setting up on a hard surface, such as crane mats.

One means of reducing the track bearing pressures calls for placement of a layer of sand or other relatively fine granular material over the mats (Fig. 1.16). Such a

Figure 1.16 Sand Layer Over Timber Crane Mats *(White Construction, Inc.)*

layer 6" (150 mm) or so in thickness will compress and allow the treads to bear over their full width and over the full (sprocket-to-sprocket) length of the track. The track pressures reasonably can be calculated based on bearing on a soft surface. This method is only practical when the crane will make only a few lifts and will not travel. Making a large number of lifts or traveling may cause the granular material to be worked out from under the tracks, thus losing its effectiveness as a load spreading layer.

The author has seen cases where a contractor has proposed laying sheets of plywood or rubber over the crane mats to obtain a soft bearing surface. This simply is not effective. The treads may rise up 0.5" (13 mm) or so from the flat center section to the edges and the difference in height of the sprockets above the rollers may exceed 1" (25 mm). Plywood, rubber, and other such materials will not compress as required to provide full area bearing of the tracks.

A second technique for reducing track bearing pressures, again in cases where the crane is not required to travel, is to block the ends of the crawler tracks. This calls for the insertion of steel plates or hardwood planks under the sprockets to fill up the gap seen in Fig. 1.8. The crane's operator's manual will specify the required blocking thicknesses. Blocking increases the track bearing length to that which is obtained on a soft surface (i.e. sprocket-to-sprocket). Blocking the tracks does not increase the bearing width, however. Thus, the track bearing pressures must be calculated using an area equal to the soft surface track bearing length d_{ls} and the hard surface track bearing width w_h.

When using software products to calculate the crawler track bearing pressures, one will find that an option for blocked tracks is generally not available. There is, however, a simple way around this problem. Use the software to calculate the track bearing pressures based on the crane bearing on a soft surface. This produces the correct length of bearing, but the pressures will be low, since they will be based on the full track pad width. Multiply these pressures by the ratio of the full track pad width to the hard surface track bearing width. This gives the correct track bearing pressures for blocked tracks. This conversion is written in equation form as follows.

$$p_{blocked} = p_{soft} \frac{w_s}{w_h} \tag{1.51}$$

where

$p_{blocked}$ = track bearing pressure for a blocked track;

p_{soft} = track bearing pressure calculated based on a soft surface; and all other terms are as previously defined.

Application of this conversion from soft surface bearing to hard surface bearing with blocked crawlers is demonstrated in Example 1-1.

EXAMPLE 1-1

Track bearing pressures imposed by a Liebherr LR1300 crane on a soft supporting surface for a particular lift are shown in Fig. 1.17. The track bearing width on a hard supporting surface, not shown in this output sheet, is 1,040 mm. Pertinent values needed to perform this calculation are as follows:

$p_{soft\ max} = 262.5$ kN/m^2 (38.07 psi)

$p_{soft\ min} = 43.1$ kN/m^2 (6.25 psi)

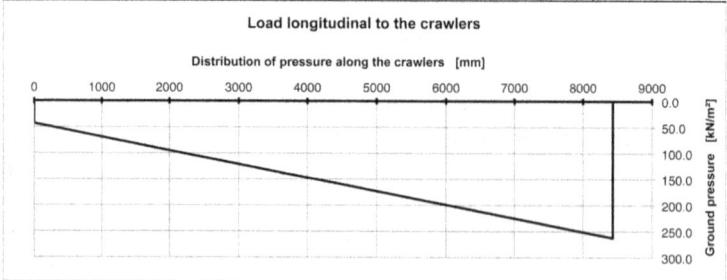

Figure 1.17 Example 1-1 Track Pressures and Dimensions

w_s = 1,200 mm (47.2 inches)

w_h = 1,040 mm (40.9 inches)

$$p_{\text{blocked max}} = p_{\text{soft max}} \frac{w_s}{w_h} = 38.07 \frac{47.2}{40.9}$$

$$p_{\text{blocked max}} = 43.9 \text{ psi}$$

$$p_{\text{blocked min}} = p_{\text{soft min}} \frac{w_s}{w_h} = 6.25 \frac{47.2}{40.9}$$

$$p_{\text{blocked min}} = 7.2 \text{ psi}$$

These values of $p_{\text{blocked max}}$ and $p_{\text{blocked min}}$ are then used with crawler track dimensions for soft bearing length d_{ls} and hard bearing width w_h in the evaluation of the supporting surface, such as crane mats.

We see in Fig. 1.17 that the track bears on the underlying surface along the full soft bearing length d_{ls}. The bearing pressure conversion of Eq. 1.51 is only valid in this case if both ends of the crawler track are blocked. That is, blocking must be installed under both the drive sprocket and the idler sprocket. However, this type of blocking installation is not always possible. In some cases, either the design of the crane or the setup location permits installing blocking under one sprocket only (Fig. 1.18). In this case, the conversion from soft surface bearing to hard surface bearing will require some additional calculation to account for the bearing length d_{lb} from the CL of a sprocket to the last track roller at the opposite end if the soft surface bearing pressure envelope extends past the last roller.

Example 1-2 demonstrates this additional calculation by taking the soft surface bearing pressure envelope shown in Fig. 1.17 and converting it to a hard surface pressure with blocking at one end only.

EXAMPLE 1-2

$p_{\text{soft max}}$ = 262.5 kN/m^2 (38.07 psi)

$p_{\text{soft min}}$ = 43.1 kN/m^2 (6.25 psi)

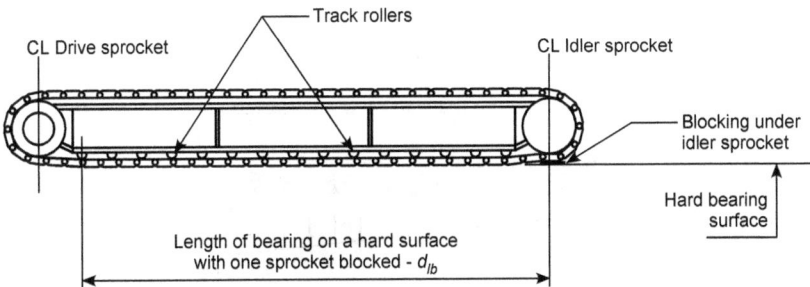

Figure 1.18 Crawler Track Bearing Length with One End Only Blocked

w_s = 1,200 mm (47.2 inches)
w_h = 1,040 mm (40.9 inches)
d_{ls} = 8,435.0 mm (332.1 inches)
d_{lh} = 7,100.0 mm (279.5 inches) (not shown in Fig. 1.17)
d_{lb} = 8,435.0 / 2 + 7,100.0 / 2 = 7,767.5 mm (305.8 inches)

$$P = \frac{(P_{soft\ max} + P_{soft\ min})}{2} w_s d_{ls} = \frac{(38.07 + 6.25)}{2} 47.2(332.1) = 347,699 \text{ pounds}$$

$$LCG = \frac{3d_{ls}P_{soft\ min} + d_{ls}(P_{soft\ max} - P_{soft\ min})}{3(P_{soft\ max} + P_{soft\ min})} = \frac{3(332.1)6.25 + 332.1(38.07 - 6.25)}{3(38.07 + 6.25)} = 126.3 \text{ in.}$$

$$e = \frac{d_{lb}}{2} - LCG = \frac{305.8}{2} - 126.3 = 26.6 \text{ inches}$$

$$p_{max} = \frac{P}{d_{lb}w_s}\left(1 + \frac{6e}{d_{lb}}\right) = \frac{347,699}{305.8(47.2)}\left[1 + \frac{6(26.6)}{305.8}\right] = 36.62 \text{ psi}$$

$$p_{max} = \frac{P}{d_{lb}w_s}\left(1 - \frac{6e}{d_{lb}}\right) = \frac{347,699}{305.8(47.2)}\left[1 - \frac{6(26.6)}{305.8}\right] = 11.51 \text{ psi}$$

$$p_{blocked\ max} = p_{soft\ max}\frac{w_s}{w_h} = 36.62\frac{47.2}{40.9} = 42.26 \text{ psi}$$

$$p_{blocked\ min} = p_{soft\ min}\frac{w_s}{w_h} = 11.51\frac{47.2}{40.9} = 13.28 \text{ psi}$$

The results are illustrated in Fig. 1.19. For reference, the maximum hard surface track pressure without any blocking is 56.29 psi. Note that the first two equations used in this example are similar in form to those discussed in Section 4.4.2 with respect to the calculation of the distribution of track bearing pressures to a longitudinally placed crane mat.

We can see from this example that the value of blocking the tracks at one end only will be minimal if the crane must swing such that the maximum track bearing

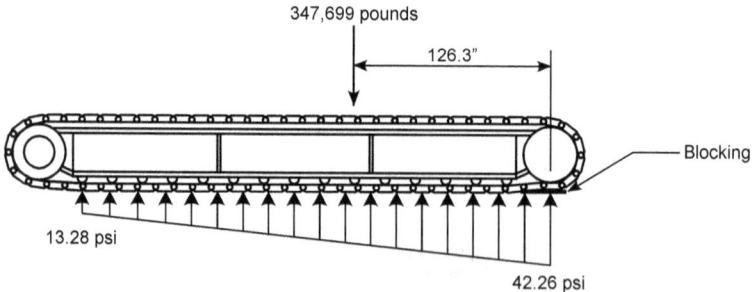

Figure 1.19 Blocked Crawler Track Bearing Pressure Calculation Results

pressures occur at the unblocked end. The advantage, or lack thereof, will become obvious when the calculations have been performed.

Last, with respect to installation of the blocking, the crane manufacturer's directions in the Operator's Manual must be followed to assure that the correct thickness of blocking is used and that the installation procedure to be used is acceptable.

1.2.4   Summary Comments on Crane Support Load Calculations

The calculation of mobile crane support loads using the methods presented in this section is not particularly complex, but it does require care to assure that correct component weight and center of gravity values are used. Further, the user must assure that the maximum outrigger load or track bearing pressure is determined by analyzing multiple cases (e.g. load at maximum radius, load at a shorter radius while swinging, minimum radius with no load on the hook, etc.).

Some readers may be surprised that this section does not include examples of the application of the equations for calculation of outrigger loads and crawler track bearing pressures. There is a reason for this. As noted at the beginning of this chapter, the need to perform these calculations has been diminishing for many years. Many crane manufacturers, as well as independent software developers, now provide applications that perform the calculation of support loads. Some of these applications are discussed in the next section. Not only do these applications perform all the "grunt work," but since they come from or are supported by the crane manufacturers, the user can reasonably assume that these applications are using all of the correct weights and dimensions and are applying the manufacturer's preferred calculation methodology. The user need only input the correct description of the crane (boom length, counterweight size, etc.), description of the lift (load and radius), and in some cases, a description of the site condition (hard or soft bearing surface).

As with all engineering calculations, there will always remain great value in understanding how a computer program solves a problem. Thus, the equations are presented, the methods of using them described, and the reader in encouraged to run through a few problems to assure understanding. However, given the present and anticipated future lack of demand on the lift planner to actually apply these calculations, the need to provide the added material to encompass examples of these methods is no longer there.

1.3   SOFTWARE FOR CRANE LOAD CALCULATIONS

Crane manufacturers have long recognized that accurate knowledge of the loads that a crane imposes on its supporting surface is critical for the safe design of the crane setup. Over the last decade or so, some of the manufacturers have made available to lift planners and crane users software that allows easy and reasonably accurate calculation of their cranes' crawler track bearing pressures or outrigger loads.

This section presents a brief discussion of some of the software products that are currently available. This is not intended to be an all-inclusive product directory (it is not), nor is it intended to be a recommendation of any particular software product.

1.3.1   Crane Manufacturer Web-Based Calculators

Some support load calculators are web-based. The user simply registers on the site and then has access to the support load calculation routines for that manufacturer's product line. Among the web-based calculators are Grove's Compu-Crane, Kobelco's Ground Pressure/Outrigger Reaction Force calculation site, and Link-Belt's Ground Bearing Information site.

In general, the site will present a list of the cranes for which support loads can be calculated. The user selects the crane model, selects various crane setup options (boom length, counterweight size, etc.) from menus, enters the load and radius into text fields, and then the software does the rest. Results can be printed or saved as a pdf file for use in a lift plan.

Web-based calculators have the obvious advantage that they are always up to date. Whenever the crane manufacturer revises the calculation routines or adds a new crane model, the changes are immediately available to the user. The only potential disadvantage is that internet access is required. However, as internet access becomes increasingly available, even in relatively remote locations, this disadvantage is disappearing.

1.3.2   Crane Manufacturer Locally Installed Software

Some support load calculators are stand-alone software products that must be installed on the user's computer. Among these products are Liebherr's LICCON lift planning package and Manitowoc's Ground Bearing Pressure Estimator (GBPE) program. These products must be procured by the user, either on digital media or by download, and installed on a computer. At present, all such locally installed executable programs are written for Microsoft Windows only.

Use of a locally installed program is generally the same as using a web-based calculator. Available crane models are shown in a menu, crane options are selected, the specifics of the lift are entered, and the support reactions are calculated. The output can be printed or saved in a variety of file formats, depending on the particular software product.

Liebherr also produces support load calculators for some crane models in the form of spreadsheet templates for Microsoft Excel. The calculation results shown in Fig. 1.17 are from such a template. These calculators can be used on any type of computer for which Excel is available (Windows and Apple Macintosh). The templates also appear to work with Apache OpenOffice, which runs under Windows, Apple's macOS, and Linux. As with the applications, the templates allow the user to select crane options and input the lifted load and radius and then the formulas in the template do the rest.

Unlike the web-based software, these products do not update automatically. Updates to the software or information for additional crane models must be obtained and installed periodically.

1.3.3   Independently Produced Software Products

There are also lift planning software products available that include support load calculation routines. Two of these are 3D Lift Plan by A1A Software, LLC (web-based) and Crane Manager by CraniMax GmbH (locally installed). The publishers of these products work with various crane manufacturers to obtain correct weight and dimension information for the crane models.

The use, advantages, and disadvantages of these independently produced products are generally the same as discussed above for the calculators produced by the crane manufacturers, with one noteworthy exception. These independently produced products may be set up with cranes from more than one manufacturer. Thus, users need only learn how to use one application to work with a variety of crane models and the output formatting will be consistent from one crane to the next.

1.4   SUPPORT LOAD PATTERNS

An obvious product of this discussion is the recognition that the location and maximum value of a crane's support loads will change as the crane performs a lift, during which the load, the radius, and/or the swing angle may change. In this section, we will examine support load patterns for crawler and truck cranes. This study will assist the reader in developing a feel for what to expect when performing support load calculations.

1.4.1   Crawler Track Bearing Pressure Patterns

Crawler track pressures will change as the crane swings, a point that should come as no surprise. Exactly *how* those pressures change is what is of interest here. We will provide one example of a crawler crane swinging with a suspended load to demonstrate this behavior. The crane used for this example is a Link-Belt Model 238 Hylab 5 fitted with the full (ABC+A) counterweight and 50 feet (15.2 m) of open throat boom. The crane is lifting a total load of 215,460 pounds (97,730 kg) at a radius of 15 feet (4.6 m). This is 90% of its rated load of 239,400 pounds (108,590 kg) for this configuration and radius.

Link-Belt's Ground Bearing Information web site calculator is used to compute the track bearing pressures through the full range of swing (0° to 360°). The maximum bearing pressures so determined are plotted in Fig. 1.20, where a swing angle of 0° is directly over the front. As expected, the lowest pressures occur when lifting directly over the side (swing angles of 90° and 270°). The maximum pressures occur when lifting over either end, but not directly over the end. Specifically, four peaks occur when the swing angle is about 33° either side of directly over the end (33°, 147°, 213°, and 327°). Due to the symmetry of the lower (x_o of Fig. 1.5 is zero), the four maximum pressure values are equal, as are the two minimum pressures.

This general pattern is typical among crawler cranes, but the specific angles at which the peak pressures occur varies. For example, some cranes exhibit peak angles at less than 20° from directly over the end. These variations depend on the size and loading of the crane, as well as the calculation method used to compute the track bearing pressures. Thus, the curve of Fig. 1.20 may be considered to be a typical behavioral pattern of a crawler crane, but the reader must recognize that each crane and each lift configuration will be unique. That is, this curve is an example, not a design tool.

1.4.2 Truck Crane Outrigger Load Patterns

As with crawler track bearing pressures, outrigger loads will similarly change as the crane swings. This behavior will also be illustrated with an example. The crane used is again a Link-Belt model, here an HC-278H II fitted with full counterweight (ABCDE+AB) and 200 feet (61 m) of open throat boom. The lift configuration for

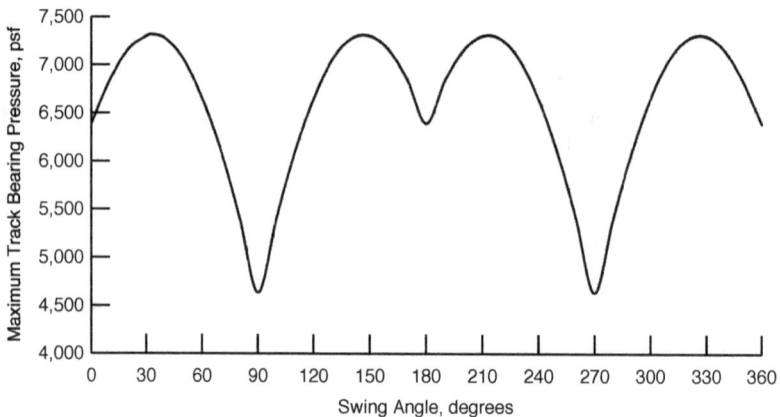

Figure 1.20 Variation of Track Bearing Pressure with Swing Angle

this example is 34,050 pounds (15,445 kg) at a radius of 100 feet (30.5 m), which is 75% of its rated load of 45,400 pounds (20,590 kg) for this configuration and radius.

Link-Belt's Ground Bearing Information web site calculator is again used to compute the outrigger loads through the range of swing angles. The maximum outrigger loads are plotted in Fig. 1.21, where a swing angle of 0° is directly over the rear. The highest outrigger loads occur when lifting over the front (this crane is equipped with a front outrigger), with the maximum outrigger load reached when lifting at a swing angle of 134°, which is 46° to the right of directly over the front. A second peak outrigger load occurs at a similar swing angle to the left of over the front. Since the outriggers are staggered longitudinally, as illustrated in Fig. 1.4, this curve is not symmetrical, as is the curve of Fig. 1.20 for the crawler crane for which the center of rotation is at the centroid of the bearing footprint.

As was noted for crawler cranes, other outrigger cranes (and this crane with a different load-radius combination) exhibit similar patterns, but with different relationships between maximum and minimum values and different swing angles at which the peaks occur. Also as with the crawler cranes, the differences in calculated outrigger loads are attributable to physical differences among the cranes and the differing results of the various outrigger load calculation methods. Further, some outrigger load calculators only give results for swing angles at 45° increments (0°, 45°, 90°, 135°, etc.). Thus, it remains the responsibility of the lift planner to understand the applied calculation methods and interpret the results rationally.

1.5 VARIABLE LOAD EFFECTS

All of the calculations of crane support loads discussed to this point are based on static weights only. While this is by far the most common approach used in these

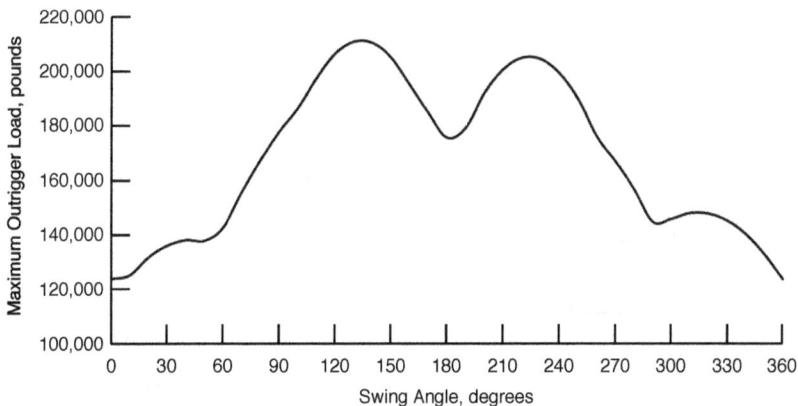

Figure 1.21 Variation of Outrigger Load with Swing Angle

calculations, some understanding should be developed with respect to the effects of other loads that may act on the crane during a lift. Although these dynamic and transient loads may not normally be significant with respect to crane support design, the lift planner/engineer should recognize their effects for that rare case in which these effects should be included in the design.

1.5.1   Effect of Change in Boom Tip Load

A crane's support loads are obviously due to a combination of the self weight of the crane and the weight of the lifted load, here considered as the load at the boom tip or jib tip. The crane's weight is fixed and usually well known. The accuracy with which the weight of the boom tip or jib tip load is known, however, varies. A change in boom tip load may be due to a small change in the lift plan, such as a change in the below-the-hook rigging, or as a result of an error in the knowledge of the weight of the payload. This presents the question: How does a change in the boom tip load translates into a change in the support reactions?

The answer to this question is: It depends. The response of the crane to a change in boom tip load is a function of the magnitude of that load relative to the weight of the crane. We will use as an example the track bearing pressures of a Manitowoc 2250 Series 3 crane. Track bearing pressures for this crane in two different configurations are plotted in Fig. 1.22. As indicated, the curve of Fig. 1.22a is for the crane fitted with a 150-foot (45.7 m) boom operating at a radius of 50 feet (15.2 m). The curve of Fig. 1.22b is for the crane fitted with a 300-foot (91.4 m) boom operating at a radius of 200 feet (61.0 m). In all cases, the crane is lifting over the front at the swing angle that produces the maximum track bearing pressure.

The short boom/short radius case presents the greater variability in bearing pressures. At low loads, the crane weight overbalances the boom tip load and the

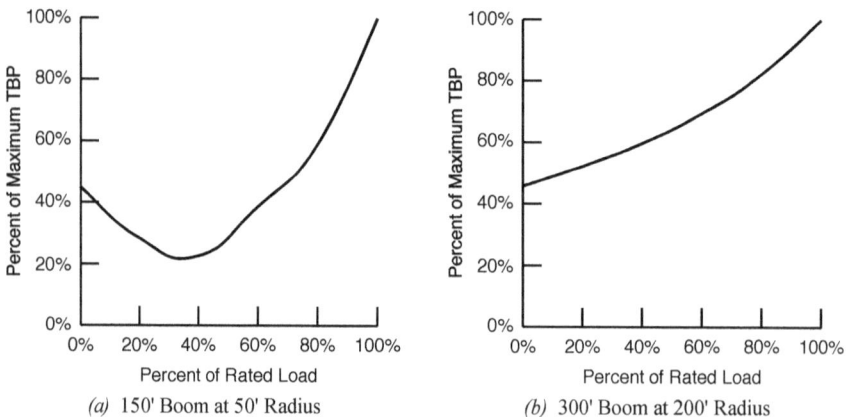

Figure 1.22 Variation of Track Bearing Pressure (TBP) with Boom Tip Load

maximum track pressure occurs at the rear of the crawler tracks. As the boom tip load is increased, the crane weight is balanced, resulting in a minimum bearing pressure. Further increases in the boom tip load result in increasing bearing pressures, now at the front of the tracks, until the maximum pressure is reached at 100% of the rated load.

The long boom/long radius case shows a markedly different track bearing pressure pattern. Here, the center of gravity of the crane's weight is to the front, so the maximum pressure with no boom tip load is at the front of the tracks and simply increases as the boom tip load increases.

Of particular significance is the way Fig. 1.22 is drawn. The boom tip load is shown as a percent of the rated load, the track bearing pressure is shown as a percent of the maximum pressure, and both axes are drawn to the same scale. Noting the slopes of the curves, we can see how a change in boom tip load equates to a change in track bearing pressure. For example, a 6.67% increase in the boom tip load from 75% of the rated load to 80% with the short boom configuration results in an 11.8% increase in the track bearing pressure (from 52.3% of the maximum value to 58.5%). The same increase in the boom tip load for the long boom configuration results in a much lower percentage increase in the track bearing pressure (about 4.5%), as indicated by the flatter slope of the curve.

Fig. 1.23 repeats this exercise for an outrigger-supported crane, here a Link-Belt HTC-8640. The maximum outrigger load was calculated for the two indicated configurations, starting with no load at the boom tip and increasing to the full rated load. The patterns shown have been observed with other outrigger-supported cranes, as well. With some cranes, however, the curve drops prior to rising, although not as pronounced as the curve of Fig. 1.22*a*. This dip in the curve was seen with a short boom at a long radius or a long boom at a short radius.

Figs. 1.22 and 1.23 are only examples based on two cranes, each in two specific configurations. The patterns of changes in support loads vs. changes in boom tip load differ from one crane to the next and, as seen here, even for one crane in

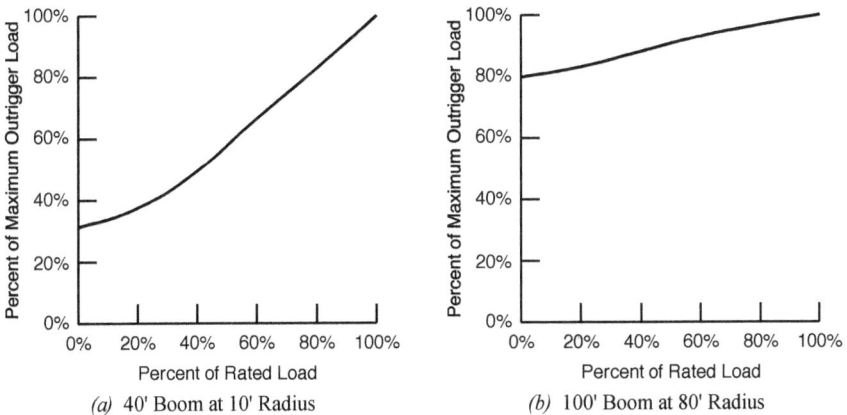

(a) 40' Boom at 10' Radius   (b) 100' Boom at 80' Radius

Figure 1.23 Variation of Outrigger Load with Boom Tip Load

different configurations. The important conclusion is that the lift planner/engineer must recognize the possible sensitivity of calculated support loads to a change in the boom tip load and work accordingly. If the weight of the payload is uncertain, use an upper bound value. Calculating the support loads for a range of payload-radius combinations can be done to develop an envelope of support load values. And as seen in Fig. 1.22, the maximum pressures may occur with no load on the hook. This case, too, must be checked.

1.5.2   Vertical Impact

Vertical impact is a dynamic force caused by the acceleration or deceleration of the lifted load (this is a hoisting function) or by a shifting of the load, as may occur when a load renders in the rigging.

A dynamic force caused by the hoisting of the lifted load should be very low. A competent crane operator will accelerate and decelerate the load slowly and otherwise avoid any type of abrupt movements. Shifting or rendering of the payload in the rigging will be minimal if the load is rigged properly. Thus, vertical impact forces will normally be trivial and need not be considered in crane support design, just as they are not considered in the structural design of the crane.

In a special situation in which vertical dynamic forces must be considered, note that an increase at the boom tip due to impact is the same as an increase due to any other change in the boom tip load, with one important difference. Dynamic loads are transient. That is, they act for a very short time, often measured only in seconds. It is this very short-lived nature of dynamic forces that diminishes the concern about vertical impact.

1.5.3   Horizontal Dynamic Forces

Horizontal forces can be imposed on a crane as a result of a number of actions. Swinging results in a dynamic force acting at the boom tip perpendicular to the boom, the interactions of the cranes in a multiple crane lift can pull the load lines out of plumb, which translates into a horizontal force, and wind acting on the boom and the suspended load creates a horizontal force.

A horizontal force acting at the boom tip creates an overturning moment that alters the support reactions at ground level. This effect can be illustrated with two very simple examples. Example 1-3 examines this effect on an outrigger-supported crane and Example 1-4 looks at a crawler crane.

EXAMPLE 1-3

Calculate the change in outrigger loads due to a horizontal force acting at the boom tip. For the purpose of this example, this analysis is performed with the crane lifting directly over the rear only. The horizontal force is acting outward in line with the operating radius.

Crane model – Grove RT9150E
Boom length = 104.5 feet
Outrigger spacing (front-to-rear) = 29 feet
Radius = 50 feet
Boom tip height = 100 feet
Rated load = 48,350 pounds
Total lifted load at boom tip = 20,000 pounds (41.4% of rated load)
Horizontal force = 1,000 pounds (2.1% of the rated load; 5% of the lifted load)

From Grove's Compu-Crane web site, we find the following outrigger loads for the specified lift:
Maximum rear outrigger load = 65,987 pounds
Minimum front outrigger load = 26,111 pounds

$$\text{Overturning moment} = 100 \times 1{,}000 = 100{,}000 \text{ pound-feet}$$

$$\text{Outrigger load change} = \frac{100{,}000}{29} = 3{,}448 \text{ pounds total} = \frac{3{,}448}{2} = 1{,}724 \text{ pounds per outrigger}$$

This outrigger load change will be positive (thus increasing the static weight outrigger loads) at the rear outriggers and negative (thus decreasing the static weight outrigger loads) at the front outriggers. Therefore, the outrigger loads due to static loading will change to the following values when the specified horizontal load at the boom tip is taken into account:

Maximum rear outrigger load = 65,987 + 1,724 = 67,711 pounds (up 2.6%)
Minimum front outrigger load = 26,111 - 1,724 = 24,387 pounds (down 6.6%)

EXAMPLE 1-4
Calculate the change in crawler track bearing pressure due to a horizontal force acting at the boom tip. For the purpose of this example, this analysis is performed with the crane lifting directly over the front only. The horizontal force is acting away from the crane in line with the operating radius.
Crane model – Kobelco SL4500
Boom length = 216.5 feet
Crawler track bearing length = 30.54 feet (9.31 meters)
Crawler track bearing width = 4.00 feet (1.22 meters) each
Radius = 60 feet
Boom tip height = 220 feet
Rated load = 177,800 pounds
Total lifted load at boom tip = 133,350 pounds (75% of rated load)
Horizontal force = 3,500 pounds (2.0% of the rated load; 2.6% of the lifted load)

From Kobelco's Ground Pressure/Outrigger Reaction Force web site, we find the following track bearing pressures for the specified lift:

Maximum track bearing pressure = 6,929 psf

Minimum track bearing pressure = 1,450 psf

Overturning moment = 220 x 3,500 = 770,000 pound-feet

$$\text{Bearing Area } S_x = \frac{2(4.00)30.54^2}{6} = 1,245 \text{ ft}^3$$

$$\text{Bearing pressure change} = \frac{770,000}{1,245} = 619 \text{ psf}$$

This change will be positive (increasing the static weight track bearing pressure) at the front end of the tracks and negative (decreasing the static weight track bearing pressure) at the rear end of the tracks. Thus, the crawler track bearing pressures due to static loading will change to the following values when the specified horizontal load at the boom tip is taken into account:

Maximum track bearing pressure = 6,929 + 619 = 7,548 psf (up 8.9%)

Minimum track bearing pressure = 1,450 - 619 = 831 psf (down 42.7%)

We see that the determination of the effect of a horizontal force acting at the boom tip on the support loads is a simple calculation. It is also rarely a necessary calculation. Mobile crane design provides for only very small transverse horizontal forces, typically only 2% of the rated load at the boom tip, so lifting operations must be planned and executed so as to minimize such horizontal forces. It follows, then, that explicit calculation of support reaction changes due to horizontal forces is generally not a normal part of crane support design. When significant horizontal loads will occur, both the crane's capacity and support design must take this into consideration.

There is also a statistical consideration. Crane supports are designed for the maximum reactions that will occur based on the planned lifts. The worst case normally occurs as the crane swings, creating a maximum outrigger load or a peak track bearing pressure. This behavior is illustrated in Figs. 1.20 and 1.21. Although the crawler crane example showed a track bearing pressure increase of almost 9% due to the horizontal load at the boom tip, the probability of the passing maximum support loading occurring simultaneously with a load increasing effect due to a transient horizontal force is relatively low. Recognition of this behavior further supports the established practice of designing crane supports based on static loads only.

There is one special case within the realm of horizontal forces that must be highlighted. That is wind load. The industry recognizes that mobile cranes must occasionally operate under windy conditions. Normal crane practice restricts lifting when the wind speed is greater than 15 to 20 miles per hour (7 to 9 meters per second), depending on the crane model and boom length. The wind pressure at such low speeds will be about 3 pounds per square foot (145 Pa). Consequently, the horizontal force due to wind is usually quite low.

There are cases, however, where lifting at higher wind speeds is necessary. Each crane manufacturer publishes special instructions for operating at higher wind speeds, usually calling for derating of the crane. In these special cases, the wind pressure can reach as high as 9 to 10 psf (430 to 480 Pa). The wind load is a function of the sail area of the lifted load and its shape (flat, cylindrical, etc.), as well as this pressure. Whether or not consideration of the wind load is necessary when evaluating the support of the crane will be based on the magnitude of the wind load and the boom tip height. These issues must be examined on a case-by-case basis. In the author's experience, the need to include wind load in the crane support calculations is extremely rare.

1.6 APPLICATION SUMMARY

This chapter presents methods for calculating the support loads, either outrigger loads or crawler track bearing pressures, that a crane imposes on its supporting surface. This closing section of the chapter summarizes the methods and equations that are to be used in the application of this material.

Two methods are developed for the calculation of outrigger loads. Both of these methods have been used in practice for many years and, based on this history, both methods may be assumed to produce acceptable results. Likewise, two methods are developed for the calculation of crawler track bearing pressures and again, both methods have been in use for many years and appear to produce acceptable results. However, the results (outrigger loads or track bearing pressures) are not identical and that can present a problem.

The author has seen some lift planners try to minimize the needed crane support work by using approaches to the calculation of crane support loads that produce "low ball" results. The most common of these is to calculate track bearing pressures based on a soft supporting surface when the crane is clearly supported on a hard surface. One can easily envision lift planners using these differing support load calculation methods to seek out the lowest values. This is obviously inappropriate at best and potentially dangerous at worst. The emphasis in the calculation of mobile crane support loads must be on safety, not on convenience or saving a few dollars.

While this chapter presents the support calculation methods in detail, the preferred approach is to use a software product provided by the crane manufacturer or a third-party vendor product that is supported by the crane manufacturer. This approach provides the best assurance that the input values are correct and that the calculation methods applied are those acceptable to the manufacturer.

With these cautionary statements in mind, the path forward requires the accumulation of all of the necessary information about the crane (weights, dimensions, configuration) and about the lift (loads, radii, and range of swing).

Multiple analyses of the crane are typically needed to be sure that the worst case loading is addressed. This may occur during the crane assembly, during a lift, or when the crane is standing by with no load on the hook. Last, the lift planner must determine if consideration of the effects of wind or other dynamic loading should be taken into account in the calculation of the support loads. In general, if dynamic loading from any source is great enough that derating of the crane becomes necessary, then consideration should be given to the effect of that dynamic loading on the design of the crane's support.

2 Strength of the Supporting Surface

Chapter 1 presents methods of calculating the loads imposed by a mobile crane on its supporting surface. Chapter 2 addresses the other side of the support equation: the strength of the surface on which the crane bears. Without a reliable understanding of the strength of the supporting surface, whether soil or some type of structure, one cannot plan a safe mobile crane installation.

The presentation in this chapter by necessity only scratches the surface of the subjects of geotechnical and structural engineering. Development of an expertise in these fields requires significant study and cannot be covered meaningfully in one chapter of one book. Rather, the goal here is to provide enough general background that the reader will understand the basic principles at work and will know what questions to ask when planning a mobile crane installation.

2.1 GEOTECHNICAL EXPLORATION

Determination of the bearing capacity of soil requires knowing the types of soil present (sand, clay, gravel, etc.), the consistency of that soil (that is, the level of compaction), and the presence of moisture in the soil, including the elevation of the water table. This is accomplished by means of a number of different methods, the most suitable of which depends on the nature of the site.

The most comprehensive and reliable means by which soil bearing capacity can be established is through the services of a geotechnical engineering firm. The geotech firm will send a crew to the site who will take one or more borings of the soil. These borings may extend dozens of feet below the surface, depending on the make-up of the soil. Soil samples will be extracted and subjected to laboratory tests to accurately assess their properties (density, shear strength, moisture content, etc.). Tests also may be made on site to further measure the soil conditions. The geotech firm ultimately will issue a detailed report in which its findings are compiled and an allowable soil bearing pressure for the location is stated.

This method of establishing an allowable soil bearing pressure is clearly the best. It is also the most expensive and the most time consuming. When making one or more major lifts in a sensitive facility, such as a refinery or a power plant, hiring a geotechnical engineering firm to perform such a study is standard practice. When making more routine lifts, the schedule and budget often cannot support this level of effort. Thus, alternative approaches must be used.

Often a good option is to seek out existing geotechnical reports from past work at or close to the location where the crane will be set up. Many owners, particularly owners of major industrial facilities, maintain records of all past work performed at the site, including past geotechnical investigations. As long as no work involving excavation has been performed in the area, a geotechnical report that is, say, five years old can still be used to gain an understanding of ground conditions at the site.

In some cases, a simpler method of exploration may be adequate. One such method is the digging of a test pit using a backhoe. As the soil below the surface is exposed, it can be visually inspected to determine the type of soil (i.e., sand, gravel, etc.). Further, hand-held tools, such as a pocket penetrometer, can be used to measure soil strength characteristics. While a test pit can be dug faster than performing a series of borings, particularly if a backhoe is readily available on site, this approach still requires a certain level of geotechnical expertise to make the right measurements and to interpret the results.

One often suggested method of evaluating soil bearing capacity is, in reality, a very poor method. This method calls for loading a small area of the ground surface in such a manner as to produce the same bearing pressure that will be imposed by the crane. The problem with this method is that the pressures imposed upon soil layers below the surface are not accurately represented.

As an example, consider the soil illustrated in Fig. 2.1. A mass of relatively soft soil is overlaid with a 3' (0.9 m) thick layer of hard material. The two loading conditions shown in the figure are examined in Example 2-1.

EXAMPLE 2-1

We will use the common approximation that the surface load spreads through the hard layer at an angle of 60° from horizontal, as shown by the dashed lines in the figure, to evaluate the soil bearing pressure at the top of the soft soil layer. (Refer to pages 58 and 59 for further discussions of this approximation.)

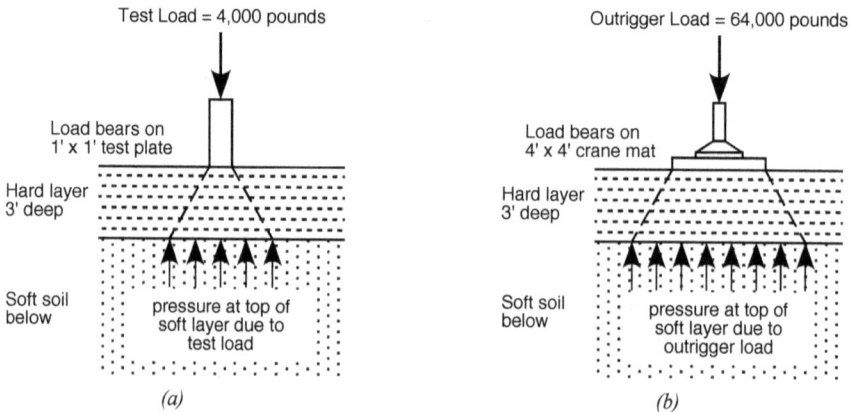

Figure 2.1 Illustration of Inaccurate Bearing Capacity Test Method

The first set of calculations determines the average bearing pressure at the top of the soft soil layer due to the test load of 4,000 pounds on a 1' x 1' plate (Fig. 2.1a).

$$q_{surface} = \frac{P}{A_{surface}} = \frac{4,000}{1.0 \times 1.0} = 4,000 \text{ psf}$$

$$A_{3\text{-feet}} = \left(1 + 2\frac{3}{\tan 60°}\right)^2 = 4.46^2 = 19.9 \text{ square feet}$$

$$q_{3\text{-feet}} = \frac{P}{A_{3\text{-feet}}} = \frac{4,000}{19.9} = 201 \text{ psf}$$

The second set of calculations determines the average bearing pressure at the top of the soft soil layer due to the outrigger load of 64,000 pounds on a 4' x 4' mat (Fig. 2.1b).

$$q_{surface} = \frac{P}{A_{surface}} = \frac{64,000}{4.0 \times 4.0} = 4,000 \text{ psf}$$

$$A_{3\text{-feet}} = \left(4 + 2\frac{3}{\tan 60°}\right)^2 = 7.46^2 = 55.7 \text{ square feet}$$

$$q_{3\text{-feet}} = \frac{P}{A_{3\text{-feet}}} = \frac{64,000}{55.7} = 1,149 \text{ psf}$$

Although the ground bearing pressure at the surface in the test is equal to the surface pressure due to the outrigger loading, the analysis shows that the average bearing pressure at the top of the soft soil layer due to the crane loading (1,149 psf) is significantly greater than the pressure that is imposed by the test load (201 psf).

This 60° approximation is just that: an approximation. A reasonable question, then, is: How realistic are the values produced using this method? We can examine these numbers by means of curves, or "pressure bulbs," that have been developed by application of the Boussinesq equations (Fig. 2.2). Similar illustrations of these curves are found in many soil mechanics and foundation design texts (e.g. Bowles 1996; NAVFAC 1986a). The curves represent the indicated ratios of vertical pressure q at the indicated location to the applied surface pressure q_a.

The curves of Fig. 2.2 are drawn based on coordinates that are multiples of the width b of the square footing. If we recast this figure in terms of the dimensions used in Example 2-1, we arrive at Fig. 2.3. Fig. 2.3a applies to the test case using a 1' square plate. We see that the peak pressure at a level three feet below the surface is somewhat greater than that found by the ratio $q / q_a = 0.05$, or 0.05 x 4,000 = 200 psf. Fig. 2.3b applies to the 4' square crane mat. Here, the peak pressure at three feet below the surface equates to $q / q_a = 0.5$, which is 0.5 x 4,000 = 2,000 psf. The pressure at the limits of a wider area (7.46 feet is considered in the 60° method) is about 0.12 x 4,000 = 480 psf, and the average pressure at this level is closer to 1,300 psf.

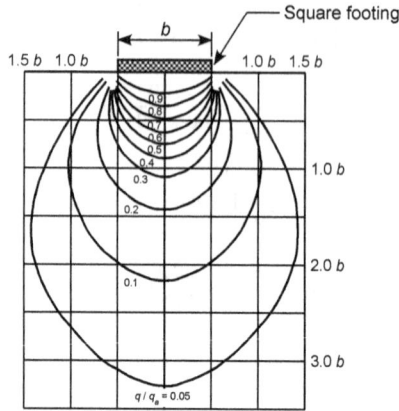

Figure 2.2 Boussinesq Pressure Bulbs for a Square Footing

Thus, although the 60° approximation method is not exact, in that it doesn't agree precisely with this other widely accepted calculation method, the results are reasonable and certainly within the degree of accuracy with which these crane support calculations are performed.

Now having looked at the comparison of the field bearing test to the actual crane load, we can see that this test, regardless of how often it may be proposed, does not stress the complete soil mass realistically and, therefore, cannot be relied upon to evaluate the soil bearing capacity.

Much can also be learned about a particular site simply by looking at the surface conditions. The following observations are often useful for evaluation of the soil.

- Signs of recent excavations and backfilling, especially near a retaining wall or the basement of a building, should be considered a red flag. If the backfilling was not placed and compacted in layers and tested for

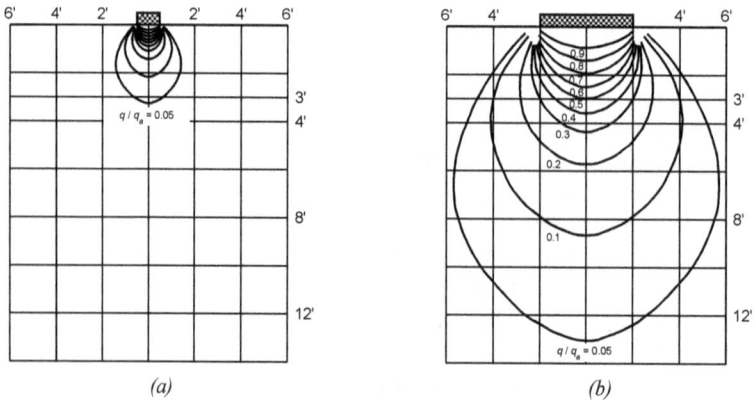

(a)   (b)

Figure 2.3 Boussinesq Pressure Bulbs for Example 2-1

compaction, the area may be of markedly lower bearing capacity than the surrounding ground.

- One should look for signs of distress in pavement when working on a street or in a parking lot. Cracked and/or sunken pavement can be a sign of a void below, indicating the need for additional exploration.
- Look for sewers, manholes, catch basins, and the like. These fixtures are always connected to underground pipes. Manhole covers should be removed to determine how deep the pipes are and where they go.
- If necessary for better accuracy in evaluating site conditions, ground penetrating radar, a magnetometer, or other similar equipment can be used to search for subsurface utilities and voids. If the crane will be set up over buried pipes or subsurface structures, additional study will be necessary, as is discussed in Section 2.4.

2.2   SOIL PROPERTIES

The only soil property that is needed for the majority of mobile crane installations is the soil bearing capacity. When subsurface pipes or other structures must be evaluated, the soil density, water table elevation and angle of internal friction are needed. Last, if the crane installation is particularly unique and demands a more rigorous soil-structure analysis, elastic properties of the soil may also be required.

2.2.1   Soil Bearing Capacity

The bearing capacity of a mass of soil is broadly a function of two characteristics of the soil. These are its strength (that is, its resistance to a stress-type failure) and its stiffness (the magnitude of deformation that will occur in response to a particular load). The strength characteristic affects the determination of bearing capacity for the design of foundations for permanent structures as well as for the design of mobile crane installations. The stiffness characteristic is always a concern in the design of permanent structures, but may not be as significant for crane installations.

In very general terms, the strength component of the bearing capacity of a mass of soil is a function of the shear strength and the cohesion of the soil. These properties are determined through the geotechnical explorations performed at the site. With this information in hand, the engineer can then use any of the many equations in the engineering literature that equate the raw soil properties to bearing capacity. The specific calculation method is dependent on the type of soil. Regardless of what is being supported (a crane, a building or anything else), the bearing capacity of the soil must be known and cannot under any circumstances be exceeded. Loading the soil beyond its bearing capacity will result in a failure of the support and in the case of a crane, overturning.

The soil mass must also be evaluated for its stiffness; that is, how the soil will deform under load. Unlike structural elements such as a steel beam, the deformation of soil has two components. These are immediate settlement, which is the deformation that occurs as soon as a load is applied, and consolidation, which is a long-term effect in which the soil compresses gradually over time. Both types of stiffness are of concern in the design of a permanent structure. Long-term consolidation is of little or no consequence with respect to the design of a mobile crane installation. Even when a mobile crane is to be used for many months in a single location, the crane operator can compensate for soil consolidation by re-leveling the crane periodically.

The soil bearing capacity as determined by the geotechnical engineer based on the soil's shear strength and cohesion is properly termed the ultimate bearing capacity. This value must be divided by a factor of safety to arrive at an allowable bearing pressure. (Some foundation design today is performed by load and resistance factor design methodologies. Mobile crane installations are generally designed using the more classical allowable strength design approach. Thus, this discussion will consistently apply allowable strength design terminology.) The allowable bearing pressure for use in the design of foundations for permanent structures is commonly based on a factor of safety of about 3. That is, the allowable bearing pressure is equal to the ultimate bearing capacity divided by 3. If the expected magnitude of soil deformation, either immediate settlement or long-term consolidation, is excessive, the allowable bearing pressure may have to be reduced even further. This practice, while appropriate for permanent structures, may be unnecessarily conservative for mobile crane installations.

The primary purpose of a factor of safety (usually called a design factor in structural engineering) is to compensate for uncertainties. The less reliably one can quantify the forces acting on a structure or the strength of that structure, the greater the uncertainty of the performance of the structure. Reliability is provided in the design by using a larger design (or safety) factor. In the case of a permanent structure, the uncertainties of loading over the life of the structure can be significant. Changes in weather patterns can alter the wind climate, changes in uses of a facility can alter the live loading, etc. Therefore, a relatively large factor of safety for the allowable soil bearing pressure is appropriate.

This uncertainty of loading is much less with a mobile crane. Even when a crane is to be set up for a lengthy construction project, the rated load of the crane effectively establishes the maximum support loads that will occur. Thus, the mobile crane's support loads to the ground surface can be fairly well defined, thereby removing some of the uncertainty in the design.

In many instances, the immediate settlement of the soil under the effect of the allowable soil bearing pressure established based on the soil strength is tolerable and, as noted above, long-term consolidation is not a concern. In such a case, the allowable bearing pressure is based on the soil strength alone. If the soil is particularly soft, though, the allowable bearing pressure may have to be reduced to limit deformation as the crane's support loads change during lifting operations.

The recognition of better defined loading and the lack of long-term consolidation as a design issue combine to provide the basis for establishing allowable soil bearing pressures for mobile crane installations using a somewhat lower factor of safety than may be used for foundation design for permanent structures. The author has often used allowable bearing pressures based on a factor of safety of 2.0 for mobile crane installation design and on a few installations that were subject to very detailed planning and review, a factor of safety of 1.5 was used (CIRIA 2003).

One last comment must be made about soil strength and allowable bearing pressure. Soil strength is negatively affected by excessive moisture. Saturated soil will fail at a lower load than will the same soil mass that has a lower, natural moisture content. This can present a problem following a heavy rain. During a storm, rain water will drain through the crane mats and soak into the soil. However, the soil will not dry out quickly after the rain stops due to the cover provided by the mats. Such a situation calls for consultation with the geotechnical engineer for evaluation of the specific conditions.

2.2.2 Use of Building Code Values

Many building codes and design standards include tables of allowable soil bearing pressures, called presumptive bearing pressures, that may be used for the design of foundations for permanent structures. These tables define allowable bearing pressures for various types of soil (sand, clay, etc.), each in a range of conditions (compact, loose, etc.). Table 2.1 is an example of such a set of presumptive bearing pressures. The values in Table 2.1 in pounds per square foot are taken from NAVFAC (1986b); the values in pascals are converted and rounded from the "pounds per square foot" source values.

The use of presumptive bearing pressures must be approached with caution. It is seen in Table 2.1 that the consistency (i.e. compactness) of the soil is extremely important in establishing the bearing capacity. Determination of the degree of compactness requires a true expertise in the field. Further, a soil mass is often not one type of soil from the surface down to a significant depth. Rather, the type of soil and its consistency may vary, perhaps greatly, in layers below the surface. A compact surface layer may disguise the reality of very soft soil below, as is discussed in conjunction with Example 2-1 and Figs. 2.1, 2.2, and 2.3.

As an example, consider the soil layers illustrated in Fig. 2.4. From a position on the ground, one would see only the very dense crushed limestone surface layer. Once a geotechnical exploration has been performed, though, one finds that the soil below is not quite as strong. Fig. 2.5 is a boring log for this soil, taken from an actual project on which the author worked a number of years ago. One convenient indicator of the soil strength that is shown in the boring log is the Standard Penetration Test blow count. We see that the top 18" (460 mm) of the layer immediately below the crushed limestone is quite strong, with blow counts of 9 to 14 blows per 6" increment. The lower layers, however, are very soft, with

TABLE 2.1 Presumptive Bearing Pressures

Type of Bearing Material	Consistency	Bearing Pressure	
		psf	Pa
Gravel, gravel-sand mixtures, boulder gravel mixtures	Very compact	14,000	670,000
	Medium to compact	10,000	480,000
	Loose	6,000	290,000
Coarse to medium sand, sand with little gravel	Very compact	8,000	380,000
	Medium to compact	6,000	290,000
	Loose	3,000	140,000
Fine to medium sand, silty or clayey medium to coarse sand	Very compact	6,000	290,000
	Medium to compact	5,000	240,000
	Loose	3,000	140,000
Homogeneous inorganic clay, sandy or silty clay	Very stiff to hard	8,000	380,000
	Medium to stiff	4,000	190,000
	Soft	1,000	50,000
Inorganic silt, sandy or clayey silt	Very stiff to hard	6,000	290,000
	Medium to stiff	3,000	140,000
	Soft	1,000	50,000

the blow counts dropping to only 4 or 5 blows per 6" increment. Thus, the overall bearing capacity of this soil mass is limited by the weaker substrata. Note that this soil structure (about 3 feet of hard material over a very soft layer) is essentially identical to the example illustrated in Fig. 2.1.

For reference, the allowable ground bearing pressure for a mobile crane installation was given in the geotechnical report from which this boring log was taken as 3,500 psf (170,000 Pa) for this location. The geotechnical engineer on this project was knowledgeable about mobile crane installation requirements and considered that in his evaluation of the soil conditions.

2.2.3   Elastic Properties of Soil

As is discussed in Chapters 4 and 5, the only soil property required by the crane mat design method developed in this handbook is the allowable ground bearing pressure. Other soil properties are not used and for most mobile crane applications, are not be needed. However, cases do arise where a more rigorous analysis of the soil and the structural elements that make up the crane support system is called for. This section provides a brief discussion of the elastic properties of soil that may be needed.

Ground surface

	Crushed limestone
1'-0"	
2'-0"	Silty clay with sand and crushed limestone (fill)
2'-6"	Medium stiff silty clay
2'-6"	Medium stiff sandy silty clay
3'-10"	Completely weathered sandstone

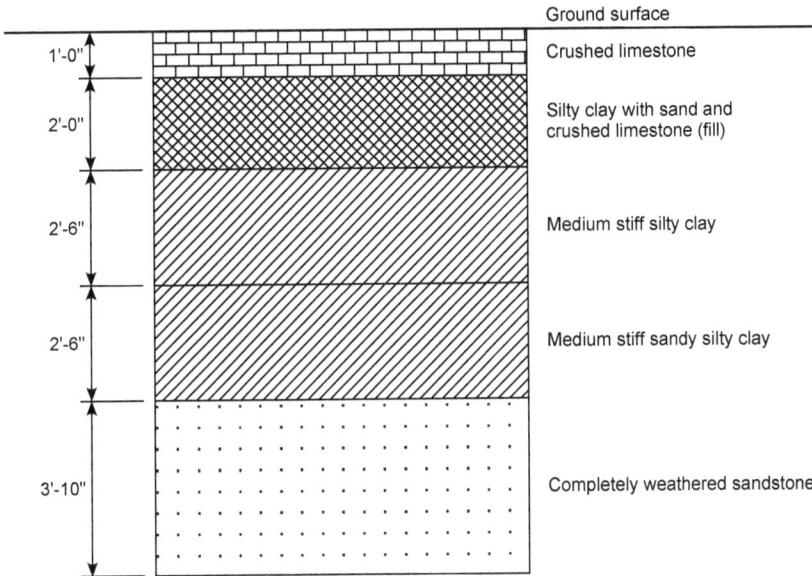

Figure 2.4 Example of Soil Layers

The most useful soil property used in soil-structure analysis is the modulus of subgrade reaction, also referred to as the coefficient of subgrade reaction, commonly indicated by the symbol k_s. This value may be thought of as a spring constant related to pressure, rather than force. The modulus of subgrade reaction is determined by a test in which a round plate up to 30" (762 mm) in diameter is loaded, usually with a hydraulic jack, so as to produce a known pressure on the surface. As the load is applied, the vertical displacement of the plate is measured. A number of such measurements are made and the resulting series of pressure-displacement points is used to construct a curve, such as the example curve shown in Fig. 2.6. The pressure q and the corresponding displacement δ are then used to calculate k_s (Eq. 2.1), typically for a value of q that corresponds to the expected service pressure or $q = q_{ult}$ (the ultimate soil bearing pressure) for failure analyses.

$$k_s = \frac{q}{\delta} \tag{2.1}$$

Values of k_s are generally expressed in units of pounds per cubic inch, kips per cubic foot, or kilonewtons per cubic meter.

Specific details of how the modulus of subgrade reaction field test is performed are defined in ASTM D1196 (ASTM 2016). Some state (in the U.S.) departments of transportation also publish their own test procedures.

As seen in the figure, the modulus of subgrade reaction is relatively linear at allowable ground bearing pressure levels, so k_s from Eq. 2.1 reasonably represents the behavior of the soil. However, as higher bearing pressures are reached, the relationship between the bearing pressure and the soil displacement becomes

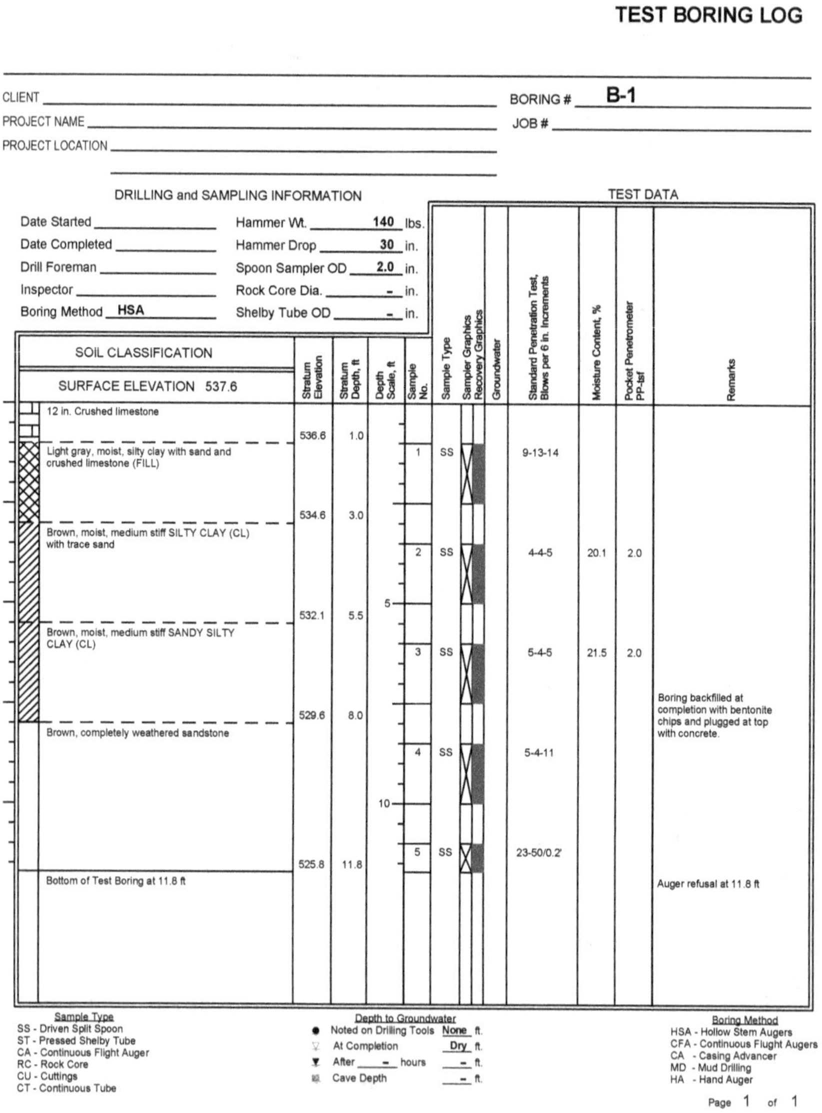

TEST BORING LOG

CLIENT _____ BORING # **B-1**

PROJECT NAME _____ JOB # _____

PROJECT LOCATION _____

DRILLING and SAMPLING INFORMATION   TEST DATA

Date Started _____ Hammer Wt. _____ **140** lbs.
Date Completed _____ Hammer Drop _____ **30** in.
Drill Foreman _____ Spoon Sampler OD __ **2.0** in.
Inspector _____ Rock Core Dia. _____ **-** in.
Boring Method **HSA**   Shelby Tube OD _____ **-** in.

SOIL CLASSIFICATION

SURFACE ELEVATION 537.6

Soil Classification	Stratum Elevation	Stratum Depth, ft	Depth Scale, ft	Sample No.	Sample Type	Sampler Graphics / Recovery Graphics	Groundwater	Standard Penetration Test, Blows per 6 in. Increments	Moisture Content, %	Pocket Penetrometer PP-tsf	Remarks
12 in. Crushed limestone											
Light gray, moist, silty clay with sand and crushed limestone (FILL)	536.6	1.0		1	SS			9-13-14			
Brown, moist, medium stiff SILTY CLAY (CL) with trace sand	534.6	3.0		2	SS			4-4-5	20.1	2.0	
Brown, moist, medium stiff SANDY SILTY CLAY (CL)	532.1	5.5	5	3	SS			5-4-5	21.5	2.0	
Brown, completely weathered sandstone	529.6	8.0		4	SS			5-4-11			Boring backfilled at completion with bentonite chips and plugged at top with concrete.
			10	5	SS			23-50/0.2'			
Bottom of Test Boring at 11.8 ft	525.8	11.8									Auger refusal at 11.8 ft

Sample Type	Depth to Groundwater	Boring Method
SS - Driven Split Spoon	● Noted on Drilling Tools **None** ft.	HSA - Hollow Stem Augers
ST - Pressed Shelby Tube	▽ At Completion **Dry** ft.	CFA - Continuous Flight Augers
CA - Continuous Flight Auger	▼ After ___ hours ___ ft.	CA - Casing Advancer
RC - Rock Core		MD - Mud Drilling
CU - Cuttings	▨ Cave Depth ___ ft.	HA - Hand Auger
CT - Continuous Tube		

Page 1 of 1

Figure 2.5 Example Soil Test Boring Log

increasingly nonlinear, with small increases in pressure resulting in large increases in displacement. Thus, k_s calculated using values of q and δ that are higher on the curve (e.g. the upper point shown in Fig. 2.6) is only a linear approximation of the actual soil behavior.

Looking back at the bearing pressure discussions in Section 2.1, we can immediately see that there is a potential weakness in the means by which the modulus of subgrade reaction is measured. The largest bearing area used in the

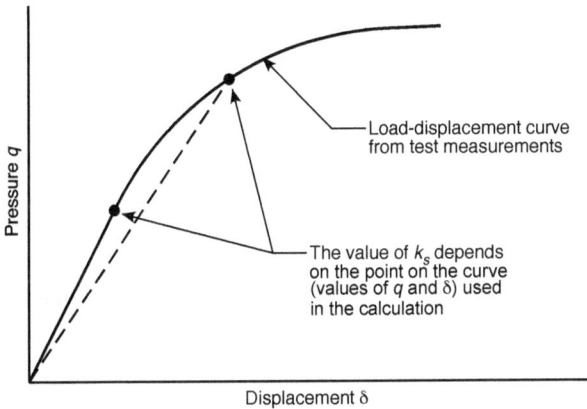

Figure 2.6 Example Modulus of Subgrade Reaction Curve from Test Data

test is that of a 30" diameter plate, which is 4.91 square feet (0.46 square meter). As illustrated in Example 2-1, the bearing pressure of, for example, 4,000 psf applied at the surface on this area will not result in the same pressure at the soft layer 3' down as will the same 4,000 psf applied on a much larger area under a crane mat. Thus, just as the determination of an acceptable bearing pressure at the surface is questionable, the determination of the modulus of subgrade reaction by loading a relatively small area at the surface is similarly questionable where the soil has layers of greatly differing properties.

Even in a reasonably homogeneous soil mass, the modulus of subgrade reaction varies as a function of the size of the foundation. Some references (e.g. Das 2016) provide the equations necessary to adjust the value of k_s to account for the difference in the test plate size relative to the foundation size. Given the practical considerations of mobile crane support engineering, however, the simple approximation of k_s given by Bowles (1996) and discussed in Section 4.2.2 often provides a suitably accurate value.

Two other elastic properties that are needed for some types of soil-foundation analysis are the modulus of elasticity E_s and Poisson's ratio μ. Determination of these properties presents the same challenges as does the determination of k_s. Unlike a manufactured material such as steel, a soil mass is very heterogeneous. This lack of consistency in the material leads to a corresponding lack of consistency in elastic properties throughout the mass.

Most soil mechanics and foundation engineering textbooks include tables of typical soil properties, including the modulus of subgrade reaction, modulus of elasticity, and Poisson's ratio. These properties are commonly shown as ranges of values for soils of various types (clay, sand, etc.) and consistencies (compact, loose, etc.). The use of values from such tables for design calculations is generally inadvisable, given the potentially great variability in the actual property values. This is another area where the services of a geotechnical engineer is needed. Field test results and textbook values will be augmented with experience-based judgment to arrive at values of the soil elastic properties that can be used reliably in an analysis of the crane support elements.

2.2.4   Other Soil Properties

There are two other properties of soil that are often useful for mobile crane support calculations. These are the soil density γ (its unit weight) and the angle of internal friction ϕ. Typical values of these two properties are given in Table 2.2 for a variety of soil types and consistencies. The values shown in the table are taken from various sources, primarily NAVFAC (1986b) and Bowles (1996). As with Table 2.1, the USCU values are from the source material and the SI values have been converted.

   The soil density γ is generally used when evaluating subsurface structures, such as buried pipes, or retaining walls. In both of these cases, the weight of the soil forms part of the loading acting on the pipe, wall, etc. The greater the soil density, the greater the load on the element will be. Therefore, when the actual soil density is uncertain, assuming a value on the high side of the likely range will produce conservative results. Any time the soil weight enters into a calculation where a greater weight is desirable, such as when designing anchors and depending on the

TABLE 2.2 Useful Typical Soil Properties

Type of Bearing Material	Consistency	Approx. Density γ		Angle of Int. Friction ϕ, deg.
		pcf	kN/m^3	
Gravel, gravel-sand mixtures, boulder gravel mixtures	Very compact	130	20.5	50
	Medium to compact	120	19.0	40
	Loose	100	15.5	35
Coarse to medium sand, sand with little gravel	Very compact	130	20.5	46
	Medium to compact	115	18.0	40
	Loose	100	15.5	34
Fine to medium sand, silty or clayey medium to coarse sand	Very compact	125	19.5	35
	Medium to compact	110	17.5	32
	Loose	105	16.5	28
Homogeneous inorganic clay, sandy or silty clay	Very stiff to hard	120	19.0	42
	Medium to stiff	105	16.5	35
	Soft	75	12.0	20
Inorganic silt, sandy or clayey silt	Very stiff to hard	120	19.0	30
	Medium to stiff	110	17.5	26
	Soft	95	15.0	20

soil for uplift resistance, then assuming a lower value of soil density will yield conservative results.

The angle of internal friction ϕ is an experimentally determined measure of the ability of soil to resist shear stress. This value is used in equations for calculating the bearing capacity of soil and for calculating the lateral earth pressure on a bulkhead, retaining wall, or other vertical structure against which a soil mass bears. If analyses that require the value of ϕ as input are to be made, obtaining the value from a geotechnical investigation is preferred over using general handbook values, such as those shown in Table 2.2. Of course, if exploration-based values are not available and the engineer must use values such as those in the table, it is prudent to investigate a proposed design using a likely range of values of ϕ to assure that an appropriately conservative solution is obtained.

2.2.5   Bearing Capacity of Frozen Soil

The strength of a soil mass is a function of the cohesion and the interlocking between the soil particles (the latter behavior is the shear strength of the soil). These characteristics are dependent upon a number of properties of the soil, including particle size and moisture content. When the temperature of the soil is such that the moisture within the mass freezes, the strength of the soil generally will increase due to this additional component of connection of the soil particles.

As an example of the change in the strength of soil at freezing temperatures, consider Fig. 2.7. The four values plotted in the figure are the results of unconfined compressive strength tests of soil at various temperatures below freezing (Shastri and Sanchez 2012). One cause of the illustrated behavior is that the water in the soil may not completely freeze at temperatures just below 32°F (0°C). This residual water will create a film that prevents full realization of the strength increase from the ice formations. This is seen in Fig. 2.7 as a small increase in strength down

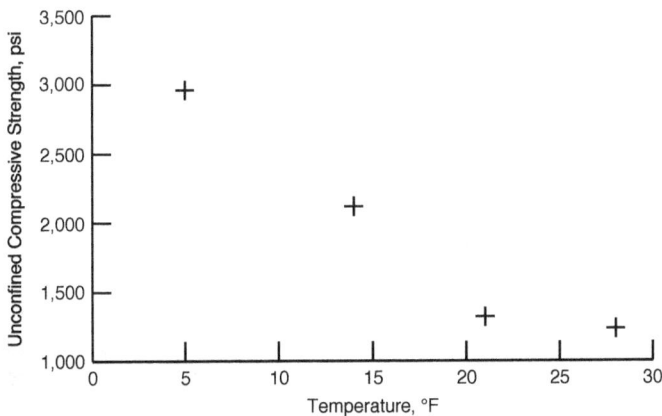

Figure 2.7 Unconfined Compressive Strength Test Results for Frozen Soil

to 21°F (-6°C), followed by a significant increase as the temperature is further reduced. Similar patterns are seen in the test results reported by Yong (1963), although the rates of strength increase with respect to temperature differ.

The values shown in Fig. 2.7 are, of course, only an example of the behavior of frozen soil. The question remains, then, of what bearing capacity can be used for low temperature crane support design. This question has no easy answer.

The bearing capacity of soil at temperatures above freezing cannot simply be multiplied by some factor to arrive at a value applicable to the frozen condition. There are too many variables, including the actual soil temperature throughout the depth of significant stress (which may be two to three times the width of the crane mat) and moisture content, to make such an approach practical. While one may reasonably assume that the soil capacity has not been reduced due to freezing, we cannot reliably assume any particular strength increase. Appropriate soil testing methods must be employed to determine the strength properties of frozen soil when an accurate assessment is needed.

The soil strength and bearing capacity issue also raises a question with respect to local hardness at the surface. That is, does frozen ground present to a crawler crane a hard surface or a soft surface? And as with the overall bearing capacity issue, there is no constant answer to this question. This local hardness at the surface will also be a function of the soil properties, moisture content and temperature, thus requiring an evaluation based on site-specific conditions and appropriate field and laboratory testing.

2.3   ADDRESSING LOW SOIL BEARING CAPACITY

As mobile cranes continue to grow in size and capacity, the demands placed on crane support design grow commensurately. Support configurations used successfully in the past are not always adequate with today's larger cranes. When the mats and/or the supporting surface are overloaded, the support design must be adjusted accordingly. The most common approach to excessive ground bearing pressure is to provide an increased bearing area. When the crane mats are overstressed, which can occur when trying to increase the bearing area, any of a number of different solutions can be employed. These include using two or more layers of mats and using mats made of materials stronger than timber, such as structural steel. The most common of these solutions are discussed in Chapter 4 as a part of the coverage of crane mat behavior analysis.

Another method of crane support improvement that must be addressed is that of increasing the soil bearing capacity. How this can be achieved depends in great part on where the weakness lies. We will consider here two options: a weak surface layer and a weak subsurface. Further, the specific improvement method used is also dependent on whether the modifications can be left in place after the lifting operations have been completed or if the site must be restored to its original condition.

When the soil bearing capacity is low due to a top layer of soft clay, organic material or other weak soils, significant improvement often can be obtained by excavating the weak material and backfilling with compacted engineered fill. The details of the work to be done (depth of excavation, type of fill to be used, amount of compaction, etc.) must be specified by a geotechnical engineer.

Another option for improving a weak soil layer calls for strengthening the soil by mechanically mixing into the soil a binding agent, such as lime, Portland cement, or fly ash. The amount of binder to be mixed into the soil must be determined based on the initial properties of the soil and the desired strength for crane support. Testing should be done to confirm that the needed strength has been achieved.

When the surface layer is reasonably strong but the subsurface is weak, as is discussed in conjunction with Figs. 2.4 and 2.5, excavation and backfilling is usually not a practical option. In this case, it may be possible to increase the allowable soil bearing pressure by building up an additional strong layer of soil. Specifically, compacted engineered fill can be built up to provide a greater thickness of sound material above the weak subsurface soil. As with the excavate-and-backfill method, the type of fill, depth and level of compaction must be specified by a geotechnical engineer.

An application of this method is illustrated in Fig. 2.8. Physical restrictions at the crane site prevented the use of longer mats under the crane, so the only path forward called for improving the soil bearing capacity. Two feet (0.6 m) of compacted crushed limestone fill was built up in the area. Three layers of timber crane mats were then placed over the fill. This arrangement reduced the bearing pressure to the weak subsurface layer to an acceptable level. The added fill was removed after the lifting operations had been completed, thus restoring the site to its original condition.

Figure 2.8 Built-Up Compacted Fill Under Crane Mats *(Senco Construction, Inc.)*

The addition of compacted fill above the native ground surface will also affect subsurface pipes and utilities. The increase in the depth of cover will reduce the live load pressure acting on the pipe or utility while increasing the dead load pressure. Evaluating the strength of subsurface pipes and utilities is discussed in the next section.

Another approach that can be used in very poor soil calls for supporting the crane structurally, rather than on the soil at the surface. Among the systems that may be used for this purpose are the following.

- Steel H piles supporting structural steel framing topped with crane mats.
- Concrete drilled shafts supporting grade beams topped with crane mats.
- Groups of micropiles (also called minipiles or pin piles) supporting structural steel framing topped with crane mats.

The structures discussed here are systems that are installed specifically for crane support. Crane installations on existing structures are discussed in Section 2.5.

The selection of the type of structural system best suited for a particular project will be driven by geotechnical conditions, the type of crane, and the required crane operations. For example, supporting a crawler crane that must travel may be best accomplished by the first option above. Steel H piles can be spaced so as to provide "runways" under each track, with the pile capacity selected to be capable of carrying the worst case track loads as the crane lifts, swings, and travels. An outrigger-supported crane may benefit from a pile group (possibly micropiles, depending on the imposed outrigger loads) under each outrigger position. Many years ago, the author utilized drilled shafts for the support of a Manitowoc 4100 Ringer® crane, using ten shafts, one under each of the ring's support crib stacks. Grade beams connecting the ten shafts and supporting conventional timber crane mats provided a good working surface for assembly of the crane.

Any of these techniques for soil improvement, any other approach to soil bearing capacity improvement, or any type of structural support for the crane may be developed through an iterative process. Site restrictions may dictate limits to crane position, crane mat size, and other aspects of support design. Arriving at an optimum solution often requires a great deal of coordination among site personnel, the lift planner/engineer, and the geotechnical engineer. The advantages and disadvantages of each element of the crane pad design must be balanced against one another until a design emerges that provides a satisfactory balance of crane support capacity and practicality.

2.4   STRUCTURES AFFECTED BY SURFACE LOADS

Soil-related structures that are affected by a mobile crane installation may be broadly divided into two groups. These are underground structures, such as buried pipes, and walls against which soil bears, such as retaining walls and bulkheads. Although not structures, embankments may be considered as being similar to retaining walls. The level of engineering effort that must be expended to investigate the effect of crane loads on each of these types of structures varies greatly, depending on the type of structure. This section provides a general overview of this subject and lists appropriate references and standards that are useful for the performance of detailed engineering of structures affected by crane loads.

2.4.1   Crane Loads Acting on Subsurface Structures

A mobile crane can develop two types of loads that may affect nearby structures. These are vertical loads and lateral earth pressures. Vertical loads in the form of ground bearing pressure at the surface affect underground pipes and subsurface structures. Vertical loads in the form of the direct crane support reactions affect structures, including slabs on grade, on which the crane bears. Lateral earth pressures, which are caused by the vertical load of the crane, act horizontally on retaining walls, bulkheads, and other similar structures.

Subsurface vertical pressures often must be calculated to evaluate buried pipes. The pressure that acts on a pipe is generally calculated as a vertical pressure or a vertical load per unit length of pipe that acts on a plane at the top of the pipe. This pressure is composed of two parts. These are the pressure due to the dead weight of the soil above the pipe and the pressure due to the weight of the mobile crane.

Consider the arrangement shown in Fig. 2.9. The ground surface will be flat for a mobile crane setup. The buried pipe has an outside diameter of D and is buried with a depth of cover H. The dashed lines indicate the approximate shape of the trench that was dug for the original installation of the pipe.

A detailed calculation of the soil dead load acting at the top of the pipe (Moser and Folkman 2008) requires knowledge of the dimensions of the trench that was dug for the original installation of the pipe, details that will rarely be known to a lifting contractor working in an existing facility. Therefore, the soil dead load is often computed using only on the pipe diameter D, the depth of cover H, and the soil density γ (e.g. AISI 1999, ALA 2005), a value sometimes called the prism load. Depending on the type of pipe and the required design method, the soil dead load may be expressed as a pressure DL_p at the top of the pipe (Eq. 2.2) or as a load per unit length DL_l of the pipe (Eq. 2.3).

$$DL_p = H\gamma \qquad\qquad (2.2)$$

$$DL_l = DH\gamma \qquad\qquad (2.3)$$

The actual soil dead load will generally be somewhat greater for rigid pipes, such as concrete pipes, installed in a trench, since soil settlement on either side of the pipe in the trench will create shear forces on either side of the soil prism directly above the pipe, thus increasing the load. Conversely, the actual soil dead

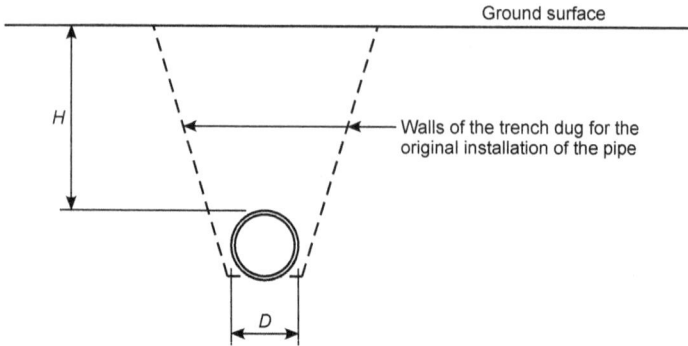

Figure 2.9 Soil Dead Load Model

load to a flexible pipe, such as a corrugated steel pipe, will be somewhat less, since the pipe will deform slightly in response to soil settlement.

The calculation of the live load due to the crane can be more complex. Each design standard for the various types of pipe specifies its own method for the calculation of the live load acting on the pipe. However, these calculation methods are all oriented toward live loads due to rubber-tired vehicles or railroad tracks. As such, they do not translate directly to subsurface pressures due to crane loads applied over a relatively large area. Further, the calculation methods applicable to pipes installed in a trench require knowledge of the original trench dimensions which, as previously noted, will rarely be known to a lifting contractor working at an existing facility. Thus, an alternate method of calculating the subsurface pressure due to crane support loads must be employed.

Recognizing the information that is typically known to a lifting contractor (or more importantly, the information that is typically not known), the simple load spreading represented by the truncated pyramid shown in Fig. 2.10 provides a practical solution (this is also the basis for the subsurface pressure calculations used in Example 2-1). Each side of the pyramid is assumed to be sloped at an angle of 60° to horizontal (Fig. 2.11). This simple geometry defines the area on a plane at the top of the pipe over which the surface load from the crane acts. Using the notation in Fig. 2.10, the length A and width B of the subsurface bearing area are defined by Eqs. 2.4 and 2.5.

$$A = a + 2\frac{H}{\tan 60°} \tag{2.4}$$

$$B = b + 2\frac{H}{\tan 60°} \tag{2.5}$$

When the pressure at the surface is uniformly distributed over the full area, as is typically the case with an outrigger load spread by a crane mat, the pressure on the subsurface bearing area is taken as uniformly distributed. When the pressure at the surface varies, as is usually the case when a crawler crane is supported

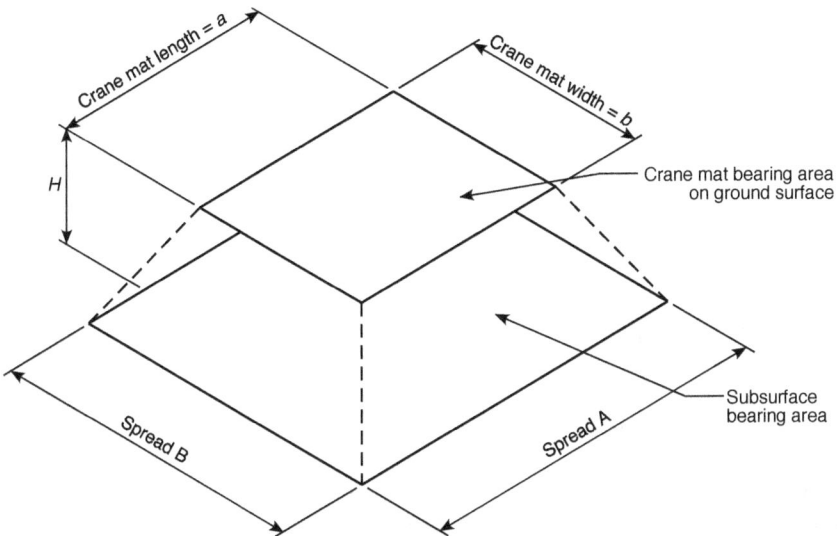

Figure 2.10 Pressure Spread Through the Truncated Pyramid

on a large crane pad, one of three approaches can be used. A simple and the most conservative approach is to treat the entire crane mat bearing area as being loaded by a pressure equal to the allowable soil bearing capacity. The second method, which is very similar, is to treat the entire crane mat bearing area as being loaded by a pressure equal to the maximum bearing pressure under the mats due to the crane loading. The third approach is to resolve the varying ground bearing pressure into a vertical load and an overturning moment and then apply that load and moment to the subsurface bearing area to arrive at a maximum subsurface crane load bearing pressure at the top of the pipe. This third approach is the most refined and, correspondingly, the most complex.

Some of the live load pressure solutions in the pipe standards (e.g. ALA 2005) apply the Boussinesq equation (Eq. 2.6) for the calculation of subsurface pressures

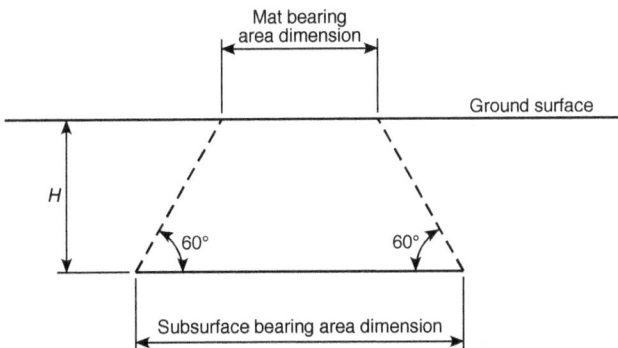

Figure 2.11 Truncated Pyramid Geometry

that are imposed on a pipe due to the live load when the live load is a concentrated load applied over a relatively small area.

$$P_p = \frac{3P_s}{2\pi H^2 \left[1+\left(\dfrac{d}{H}\right)^2\right]^{2.5}} \tag{2.6}$$

where

P_p	=	unit pressure at the plane at the top of the pipe;
P_s	=	concentrated load at the surface;
H	=	height of cover above the pipe; and,
d	=	horizontal distance from load P_s to the point at which P_p is to be determined.

While this approach is reasonable when the surface loads are concentrated loads applied on small areas, as is the case with pneumatic tires, use of the Boussinesq equation is questionable when the surface load is applied over a large area through a relatively rigid structure. Calculation of the subsurface pressure distribution from a crane by dividing the bearing area into small pieces, one foot square for example, and then calculating P_p from each discreet P_s load results in an apparent high pressure under the center of the crane area, tapering off to much lower pressures out toward the edges. This behavior could only occur if the ground surface could deform similarly to accommodate the much larger soil displacement that would occur under the center of the crane. Given the limited deflection of crane mats designed in accordance with the method developed in Chapter 4, this stiffness will result in a more uniform pressure distribution through the soil mass. Thus, the truncated pyramid pressure distribution of Figs. 2.10 and 2.11 and Eqs. 2.4 and 2.5 holds as a practical method of calculating subsurface pressures for the evaluation of buried pipes and other structures.

2.4.2   Strength of Buried Pipes

Calculation methods used to determine the strength of and effect of surface loads on different types of buried pipes are addressed by pipe-specific standards and industry guides. Following are general discussions of the design methods applicable to plain steel, concrete, and corrugated steel pipes, which are the most common types of buried pipes that may be encountered. Appropriate standards or guides must be followed for types of pipes not addressed here.

Steel pipe. Standards that address the design and installation of a buried steel pipe include ALA (2005) and AWWA (2017). These pipes often carry pressurized fluid, so the design requirements address the stresses caused by the internal pressure as well as the external loads. One can immediately see that the stresses due to the internal fluid pressure are opposite in sense to the stresses caused by the external loads from the soil. Thus, an analysis of a buried steel pipe subjected to crane loading should be performed assuming no internal pressure to assure that the worst case condition is addressed. Except where otherwise noted, the following procedure is based on the provisions of ALA (2005).

The vertical pressure above the pipe due to the soil DL_p is to be calculated as the weight of the prism of soil directly above the pipe, as given by Eq. 2.2. ALA (2005) contains tables of live load pressures to the pipe due to various highway, railway, and airport runway loads and suggests the use of the Boussinesq equation (Eq. 2.6) for the calculation of pressures due to concentrated loads. Calculation of the pressure at the pipe due to a surface pressure over a relatively large area is not addressed. Therefore, the (upper bound) vertical pressure above the pipe due to the crane load at the surface CL_p can be calculated using Eq. 2.7.

$$CL_p = \frac{pab}{AB} \tag{2.7}$$

where

p	=	ground surface pressure due to the crane;
a	=	crane mat length (Fig. 2.10);
b	=	crane mat width (Fig. 2.10);
A	=	pressure area dimension at top of pipe, calculated with Eq. 2.4; and,
B	=	pressure area dimension at top of pipe, calculated with Eq. 2.5.

A buried steel pipe is considered to be a flexible pipe in that it will exhibit relatively significant deformation due to vertical loading. Specifically, the pipe will deform into an oval shape due to the vertical pressure from the soil weight and the applied surface load. This ovality, taken as the ratio of vertical deflection to outside diameter, is calculated with Eq. 2.8.

$$\frac{\Delta y}{D} = \frac{D_1 KP}{\dfrac{(EI)_{eq}}{R^3} + 0.061E'} \tag{2.8}$$

where

Δy	=	vertical deflection of the pipe, inches (mm);
D	=	pipe outside diameter, inches (mm);
D_1	=	deflection lag factor (may be taken as 1.5);
K	=	bedding constant (may be taken as 0.1);
P	=	total pressure at the top of the pipe, psi (MPa);
	=	$DL_p + CL_p$

$(EI)_{eq}$ = equivalent pipe wall stiffness per inch (mm) of pipe length, pound-inch (N-mm);

R = pipe outside radius, inches (mm); and,

E' = modulus of soil reaction (Table 2.3), psi (MPa).

Note that one MPa, most commonly identified as 10^6 newtons per square meter, also is equal to one newton per square millimeter. Although Eq. 2.8 and the following equations are dimensionally independent, recommended units for the various quantities are shown in the lists of terms to provide consistency with the units of E' shown in Table 2.3.

The equivalent pipe wall stiffness $(EI)_{eq}$ is the sum of the stiffnesses EI of the bare pipe, the pipe lining, if any, and the external coating, if any. We can see in Eq. 2.8 that a smaller value of the equivalent pipe stiffness will result in a larger, and therefore more conservative, value of the pipe ovality, so use of the stiffness of the bare pipe is reasonable when the properties of any lining or coating are not known.

The modulus of soil reaction E' is a measure of the elastic lateral stiffness of the soil. As such, it is used in Eq. 2.8 to model the support provided to the sides of the pipe by the soil. This quantity is not the same as the modulus of subgrade reaction discussed starting on page 49. Further, it is not a value that is normally measured in a geotechnical investigation, so the value of E' often must be estimated. Table 2.3 (based on Moser and Folkman 2008) shows average values of the modulus of soil reaction for a variety of types of soil at a range of degrees of compaction. Note that the soil characteristics to be used when entering Table 2.3 are those for the soil layer in which the pipe is installed.

The ovality of the pipe from Eq. 2.8 is used to calculate the through-wall bending stress σ_{bw} (Eq. 2.9).

$$\sigma_{bw} = 4E\left(\frac{\Delta y}{D}\right)\left(\frac{t}{D}\right) \tag{2.9}$$

where

E = modulus of elasticity of the pipe wall;

t = pipe wall thickness; and,

all other terms are as previously defined.

ALA (2005) suggests an allowable bending stress equal to 0.5 F_y. Given the reduced uncertainties in loading associated with a mobile crane installation, we may consider using a higher allowable stress for the evaluation of a pipe loaded by crane reactions. An allowable bending stress of 0.75 F_y is suggested for crane installation engineering. This allowable bending stress provides a design factor of 1 / 0.75 = 1.33 with respect to outer fiber yield and 1 / 0.75 x 1.50 = 2.00 with respect to formation of a plastic moment in the pipe wall (the plastic modulus of a rectangular section is 1.50 times its section modulus).

TABLE 2.3 Average Values of Modulus of Soil Reaction E' (based on Moser and Folkman 2008)

Soil Type	E' based on Degree of Compaction of Bedding, psi (MPa)			
	Dumped	Slight, <85% Proctor, <40% relative density	Moderate, 85% – 95% Proctor, 40% – 70% relative density	High, >95% Proctor, >70% relative density
Fine-grained soils (liquid limit > 50) Soils with medium to high plasticity Inorganic silts and clays		No data available. Consult a geotechnical engineer or use $E' = 0$		
Fine-grained soils (liquid limit < 50) Soils with medium to no plasticity Inorganic silts and clays with less than 25% coarse-grained particles	50 (0.35)	200 (1.4)	400 (2.8)	1,000 (7)
Fine-grained soils (liquid limit < 50) Soils with medium to no plasticity Inorganic silts and clays with more than 25% coarse-grained particles Coarse-grained soils with fines Gravel-sand-silt mixtures containing more than 12% fines	100 (0.70)	400 (2.8)	1,000 (7)	2,000 (14)
Coarse-grained soils with little or no fines Gravels and gravel-sand mixtures containing less than 12% fines	200 (1.4)	1,000 (7)	2,000 (14)	3,000 (21)
Crushed rock	1,000 (7)	3,000 (21)	3,000 (21)	3,000 (21)
Accuracy in terms of percentage of deflection	±2	±2	±1	±0.5

The last limit state to be considered is ring buckling of the pipe. The maximum allowable total pressure P_c that may act at the top of the pipe is equal to the ring buckling pressure divided by an appropriate factor of safety. P_c is defined by Eqs. 2.10 and 2.11.

$$P_c = \frac{1}{FS} \sqrt{32 R_w B' E' \frac{(EI)_{eq}}{D^3}} \tag{2.10}$$

$$B' = \frac{1}{1 + 4e^{\left(-0.065 H/D\right)}} \tag{2.11}$$

where
FS	=	factor of safety;
	=	2.5 when $H/D \geq 2$ (as recommended in ALA 2005)
	=	3.0 when $H/D < 2$ (as recommended in ALA 2005)
R_w	=	water buoyancy factor;
	=	$1 - 0.33 \, h_w / H$
h_w	=	height of water surface above top of pipe; and,
		all other terms are as previously defined.

Application of this analysis method can be demonstrated in an example problem based on an outrigger supported on a 4' x 12' mat located directly above a subsurface steel pipe.

EXAMPLE 2-2

    Mat bearing length a = 12'-0"
    Mat bearing width b = 4'-0"
    Outrigger load = 192,000 pounds, so pressure p = 192,000 / (12 x 4) = 4,000 psf

    Pipe outside diameter D = 18"
    Pipe wall thickness t = 0.375"
    Pipe yield stress F_y = 36,000 psi
    Cover H = 3.00 feet

    Soil density γ = 110 pounds per cubic foot
    Modulus of soil reaction E' = 1,000 psi
    Water table is below the pipe, so R_w = 1.0

    The soil pressure DL_p is calculated with Eq. 2.2.

$$DL_p = H\gamma = 3.0 \times 110 = 330 \text{ psf}$$

The length and width of the bearing area at the top of the pipe are calculated with Eqs. 2.4 and 2.5. The pressure due to the crane load CL_p is then calculated with Eq. 2.7.

$$A = a + 2\frac{H}{\tan 60°} = 12.0 + 2\frac{3.0}{\tan 60°} = 15.46 \text{ feet}$$

$$B = b + 2\frac{H}{\tan 60°} = 4.0 + 2\frac{3.0}{\tan 60°} = 7.46 \text{ feet}$$

$$CL_p = \frac{pab}{AB} = \frac{4,000 \times 12.00 \times 4.00}{15.46 \times 7.46} = 1,663 \text{ psf}$$

Thus, the total pressure P on the plane at the top of the pipe is $330 + 1,663 = 1,993$ psf $= 13.8$ psi.

The ovality of the pipe and the resulting through-thickness bending stress are calculated with Eqs. 2.8 and 2.9. This requires first calculating the moment of inertia I of the pipe wall in the through thickness direction. The resulting bending stress is compared to the allowable bending stress of 0.75 $F_y = 27,000$ psi.

$$I = \frac{0.375^3}{12} = 0.0044 \text{ in.}^3$$

$$\frac{\Delta y}{D} = \frac{D_l KP}{\frac{(EI)_{eq}}{R^3} + 0.061 E'} = \frac{1.5 \times 0.1 \times 13.8}{\frac{29,000,000 \times 0.0044}{9^3} + 0.061 \times 1,000} = 0.009$$

$$\sigma_{bw} = 4E\left(\frac{\Delta y}{D}\right)\left(\frac{t}{D}\right) = 4 \times 29,000,000 \times 0.009\frac{0.375}{18.000} = 21,280 \text{ psi}$$

$$\text{Bending S.R.} = \frac{21,280}{27,000} = 0.79 < 1.00 \text{ O.K.}$$

The bending stress in the pipe wall is found to be acceptable for the defined conditions and crane loading.

Ring buckling is checked by means of Eqs. 2.10 and 2.11. The ratio of cover H to pipe diameter D is equal to 2.0, so the factor of safety used in Eq. 2.10 is taken as 2.5.

$$B' = \frac{1}{1 + 4e^{(-0.065 H/D)}} = \frac{1}{1 + 4e^{(-0.065 \times 36/18)}} = 0.222$$

$$P_c = \frac{1}{FS}\sqrt{32 R_w B' E' \frac{(EI)_{eq}}{D^3}} = \frac{1}{2.5}\sqrt{32 \times 1.0 \times 0.222 \times 1,000\frac{29,000,000 \times 0.0044}{18^3}} = 157 \text{ psi}$$

$$\text{Buckling S.R.} = \frac{13.8}{157} = 0.09 < 1.00 \text{ O.K.}$$

The total applied pressure is less than the critical ring buckling pressure, so the pipe is again found to be acceptable for the defined conditions and crane loading.

Concrete Pipe. The evaluation of a subsurface concrete pipe is addressed in ACPA (2011). Much of this manual covers the design of the pipe installation, including the type and shape of trench and the bedding of the pipe (that is, the preparation of the bottom of the trench on which the pipe is laid). The manual also defines methods for calculating the loads imposed on the pipe, but then directs the user to the ASTM standards listed in Table 2.4 for the values of the pipe strength. The full citation for each standard appears in the References section at the end of this chapter.

Concrete pipes are considered to be rigid pipes, in that the deformation of the initially circular cross section is relatively small. Consequently, the load to the top of the pipe due to the dead weight of the soil is greater than that used in the design of flexible pipes, as is discussed above in conjunction with Eqs. 2.2 and 2.3. ACPA (2011) gives an equation for the calculation of the dead load from the soil above the pipe in which the height of the soil prism is somewhat greater than simply H, as is used in Eqs. 2.2 and 2.3. Further, the soil load is increased by a Vertical Arching Factor that is a function of the type of installation that was employed when the pipe was originally laid. As previously noted, a lifting contractor working in an existing facility will rarely know such details. The maximum Vertical Arching Factor value given in ACPA (2011) is 1.45. In the absence of better information, this value should be used for the calculation of the soil dead load on a concrete pipe. Thus, the soil dead load for the analysis of a buried concrete pipe, taking into account the ACPA (2011) method of calculating the prism load and the maximum Vertical Arching Factor of 1.45, can be written as Eq. 2.12.

$$DL_l = 1.45\gamma \left[H + \frac{D(4-\pi)}{8} \right] D \qquad (2.12)$$

where
D =   outside diameter of round pipe; or,
         =   outside horizontal dimension of arch or elliptical pipe; and,
            all other terms are as previously defined.

The calculation of the pressure to the top of the pipe due to live loading in ACPA (2011) is again based on the wheel loads from rubber-tired highway vehicles. The

TABLE 2.4 ASTM Standards for Concrete Pipe Strength

Designation	Applicability
C14 / C14M	Nonreinforced concrete pipe
C76 / C76M	Reinforced concrete culvert, storm drain, and sewer pipe
C506 / C506M	Reinforced concrete arch culvert, storm drain, and sewer pipe
C507 / C507M	Reinforced concrete elliptical culvert, storm drain, and sewer pipe
C655 / C655M	Reinforced concrete culvert, storm drain, and sewer pipe
C985 / C985M	Nonreinforced concrete culvert, storm drain, and sewer pipe

spread of the load through the soil cover is based on the truncated pyramid model, with varying factors depending on the type of soil (granular or "other") and the direction of the long side of the tire footprint rectangle relative to the length of the pipe. The calculation of vertical pressure due to the crane load at the ground surface may be calculated using the truncated pyramid model recommended for the design of steel pipe analysis (Eqs. 2.4, 2.5, and 2.7, based on Figs. 2.10 and 2.11). The crane load pressure CL_p must then be converted to a load per unit length CL_l using Eq. 2.13.

$$CL_l = CL_p D \tag{2.13}$$

The crane load CL_l at the top of the pipe is assumed to be distributed over an effective supporting length of pipe L_e. This length is illustrated in Fig. 2.12 and is calculated with Eq. 2.14. This value is then used in Eq. 2.15 to calculate the crane load per unit length of pipe W_L that will be used to evaluate the strength of the pipe (ACPA 2011).

$$L_e = L + 1.75(0.75R_o) \tag{2.14}$$

$$W_L = CL_l \frac{L}{L_e} \tag{2.15}$$

where
L =   length of bearing pressure plane at the top of the pipe;
R_o =   outside diameter of round pipe; or,
     =   outside vertical dimension of arch or elliptical pipe; and,
        all other terms are as previously defined.

The length L is equal to either spread A or spread B of the crane load bearing area

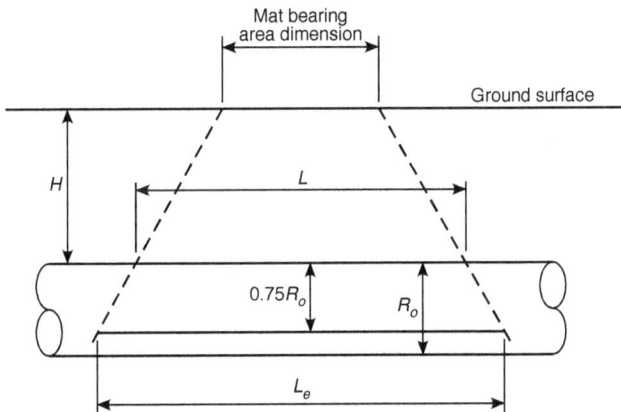

Figure 2.12 Effective Supporting Length of Concrete Pipe

(Fig. 2.10), whichever is parallel to the length of the pipe. If the pipe orientation is not known, use of the larger value will yield a conservative result.

The next value required is the bedding factor B_f. The bedding factor is based on the design and geometry of the pipe installation and, as with the trench dimensions previously discussed, may not be known to a lifting contractor. Bedding factor values for the various types of pipe installations are given in ACPA (2011). When the installation details are known, the appropriate value can be selected. When these details are not known, using a lower value of B_f produces a conservative result.

The strength of a reinforced concrete pipe is given in the ASTM standards listed in Table 2.4 as a set of D-loads, where the D-load is that load in pounds per linear foot per foot of diameter (newtons per linear meter per millimeter of diameter) that will produce a crack having a width of 0.01 inch (0.3 mm) throughout a length of at least one foot (305 mm). The required D-load ($D_{0.01}$) for the pipe based on the type of installation and the imposed loading is calculated using Eq. 2.16.

$$D_{0.01} = \left[\left(\frac{DL_l + W_F}{B_f} \right) + \frac{W_L}{B_{fLL}} \right] \frac{FS}{D_i} \qquad (2.16)$$

where

DL_l	=	soil dead load from Eq. 2.12, pounds per foot (newtons per meter);
W_F	=	weight of fluid in the pipe, pounds per foot (newtons per meter);
B_f	=	soil load bedding factor;
W_L	=	crane load from Eq. 2.15, pounds per foot (newtons per meter);
B_{fLL}	=	live load bedding factor;
FS	=	factor of safety
	=	1.00 when using the D-load values from the ASTM standards; and,
D_i	=	inside diameter of a circular pipe or inside span of an elliptical or arch pipe, feet (mm).

Given the uncertainties surrounding the determination of the bedding factor for pipes in existing facilities, the soil load and live load bedding factors can be assumed to be the same value.

Once the required D-load $D_{0.01}$ has been determined using Eq. 2.16 and the actual D-load for the pipe has been determined from the appropriate ASTM standard, comparison of the two values is all that is required to determine the adequacy (or lack thereof) of the buried pipe (that is, the D-load from the ASTM standard is an allowable load).

As with the steel pipe, application of this analysis method can be demonstrated in an example problem based on a crawler crane located directly above a subsurface concrete storm drain pipe.

EXAMPLE 2-3

Mat bearing length a = 30'-0"

Mat bearing width b = 28'-0"
Maximum ground bearing pressure p = 3,000 psf

Class IV reinforced concrete pipe with Wall B (ASTM C76 applies)
Pipe inside diameter D_i = 36"
Pipe wall thickness t = 4"
D-load to produce a 0.01-inch crack = 2,000 pounds per linear foot per foot of diameter
Cover H = 3.00 feet

Soil density γ = 110 pounds per cubic foot
Bedding and other installation details are not known reliably - use B_f = 1.7
Water table is below the pipe

The outside diameter of the pipe D = (36 + 2 x 4) / 12 = 3.67 feet.
A storm drain pipe will typically be empty, except during heavy rain. Therefore, the pipe is assumed to be empty during crane operations and the fluid load W_F is zero.

The soil pressure DL_l is calculated with Eq. 2.12.

$$DL_l = 1.45(110)\left[3.00 + \frac{3.67(4-\pi)}{8}\right]3.67 = 1,985 \text{ plf}$$

The length and width of the bearing area at the top of the pipe are calculated with Eqs. 2.4 and 2.5. The pressure due to the crane load CL_p is then calculated with Eq. 2.7.

$$A = a + 2\frac{H}{\tan 60°} = 30.0 + 2\frac{3.0}{\tan 60°} = 33.46 \text{ feet}$$

$$B = b + 2\frac{H}{\tan 60°} = 28.0 + 2\frac{3.0}{\tan 60°} = 31.46 \text{ feet}$$

$$CL_p = \frac{pab}{AB} = \frac{3,000 \times 30.00 \times 28.00}{33.46 \times 31.46} = 2,393 \text{ psf}$$

The pressure CL_p is converted to the crane load at the top of the pipe as a load per unit length using Eq. 2.13.

$$CL_l = 2,393 \times 3.67 = 8,776 \text{ pounds per foot}$$

The effective supporting length of the pipe L_e (shown in Fig. 2.12) is calculated using Eq. 2.14 and then the crane load to the pipe W_L as a load per unit length is calculated using Eq. 2.15.

$$L_e = 33.46 + 1.75(0.75 \times 3.67) = 38.28 \text{ feet}$$

$$W_L = 8,776\frac{33.46}{38.28} = 7,672 \text{ pounds per foot}$$

The load quantities now in hand are combined in Eq. 2.16 to determine the required pipe D-load to safely support the imposed loads.

$$D_{0.01} = \left[\left(\frac{DL_I + W_F}{B_f}\right) + \frac{W_L}{B_{fLL}}\right]\frac{FS}{D_i} = \left[\left(\frac{1,985 + 0}{1.7}\right) + \frac{7,672}{1.7}\right]\frac{1.00}{3.00} = 1,894 \text{ pounds per foot per foot}$$

$$\text{Utilization Ratio} = \frac{1,894}{2,000} = 0.95 < 1.00 \text{ O.K.}$$

$D_{0.01}$ is less than the D-load for this size and grade of pipe, so the pipe is found to be acceptable for the defined conditions and crane loading.

The analysis of nonreinforced concrete pipe is based on the three-edge-bearing strength given in the ASTM standards (shown in Table 2.4) in units of pounds per foot (newtons per meter), rather than the D-load. The required three-edge-bearing load (T.E.B.) is calculated using Eq. 2.17, in which all terms are as previously defined.

$$\text{T.E.B.} = \left[\left(\frac{DL_I + W_F}{B_f}\right) + \frac{W_L}{B_{fLL}}\right]FS \qquad (2.17)$$

Corrugated Steel Pipe. The design and analysis of corrugated steel pipe is addressed in AISI (1999). Like the plain steel pipes previously discussed, corrugated steel pipes are treated as flexible pipes, so the calculation of the imposed loads is very similar to that defined for plain steel pipes. And as with the other types of pipes, AISI (1999) provides live load calculation methods primarily for highway and rail service.

The soil dead load DL_p to a corrugated steel pipe is calculated as the prism load above the pipe using Eq. 2.2. The crane load CL_p to the pipe is calculated based on the truncated pyramid model using Eqs. 2.4, 2.5, and 2.7. These two pressures are combined using Eq. 2.18.

$$P = K\left(DL_p + CL_p\right) \qquad (2.18)$$

where

K = 0.86 if $H \geq D$ and Standard Density $\geq 85\%$;
 = 0.75 if $H \geq D$ and Standard Density $\geq 90\%$;
 = 0.65 if $H \geq D$ and Standard Density $\geq 95\%$;
 = 1.00 for all other conditions;
D = outside diameter of round pipe; or,
 = outside horizontal dimension of arch or elliptical pipe; and,
   all other terms are as previously defined.

Corrugated pipe is evaluated on the basis of the compressive thrust in the walls C, as calculated with Eq. 2.19. This thrust force is reduced to a compressive stress using Eq. 2.20 and the allowable compressive stress is calculated using Eqs. 2.21 through 2.23. The equations in AISI (1999) have been modified here to be dimensionally independent.

$$C = P\frac{D}{2} \tag{2.19}$$

$$f_c = \frac{C}{A} \tag{2.20}$$

When $D/r \le 294$
$$F_c = \frac{F_y}{2} \tag{2.21}$$

When $294 < D/r \le 500$
$$F_c = \left[1.212F_y - \frac{E}{3.58 \times 10^8}\left(\frac{D}{r}\right)^2\right]\frac{1}{2} \tag{2.22}$$

When $D/r > 500$
$$F_c = \frac{170E}{2\left(\dfrac{D}{r}\right)^2} \tag{2.23}$$

where

A = pipe wall area per unit length;
F_y = pipe steel yield stress (typically 33,000 psi, 230 MPa);
E = modulus of elasticity (29,000,000 psi, 200,000 MPa); and,
r = radius of gyration of the pipe wall.

The dimensions and section properties of standard corrugated pipe products are tabulated in AISI (1999).

As with the steel and concrete pipes, application of this analysis method can be demonstrated in an example problem. Again, we will use a crawler crane located directly above a subsurface corrugated steel storm drain pipe.

EXAMPLE 2-4

    Mat bearing length a = 30'-0"
    Mat bearing width b = 28'-0"
    Maximum ground bearing pressure p = 3,000 psf

    Pipe diameter D = 60"
    Pipe wall thickness = 0.064"
    Pipe corrugation profile - 2 2/3" x 1/2"
    Pipe wall area A = 0.775 in.[2] per foot (from AISI 1999, Table 7.2)

Pipe wall moment of inertia $I = 0.0227$ in.[4] per foot (from AISI 1999, Table 7.2)
Cover $H = 5.00$ feet

Soil density $\gamma = 110$ pounds per cubic foot
Compaction = 90% of Standard Proctor
Water table is below the pipe

The soil pressure DL_l is calculated with Eq. 2.2.

$$DL_p = H\gamma = 5 \times 110 = 550 \text{ psf}$$

The length and width of the bearing area at the top of the pipe are calculated with Eqs. 2.4 and 2.5. The pressure due to the crane load CL_p is then calculated with Eq. 2.7.

$$A = a + 2\frac{H}{\tan 60°} = 30.0 + 2\frac{5.0}{\tan 60°} = 35.77 \text{ feet}$$

$$B = b + 2\frac{H}{\tan 60°} = 28.0 + 2\frac{5.0}{\tan 60°} = 33.77 \text{ feet}$$

$$CL_p = \frac{pab}{AB} = \frac{3,000 \times 30.00 \times 28.00}{35.77 \times 33.77} = 2,086 \text{ psf}$$

The soil load and crane load pressures DL_p and CL_p are converted to a total effective pressure P at the top of the pipe using Eq. 2.18. Since $H = 5.00$ feet $= D = 5.00$ feet and the soil compaction is 90%, $K = 0.75$.

$$P = K\left(DL_p + CL_p\right) = 0.75\left(550 + 2,086\right) = 1,977 \text{ psf}$$

The compressive thrust C in the pipe walls is calculated using Eq. 2.19 and the corresponding compressive stress f_c is calculated using Eq. 2.20.

$$C = P\frac{D}{2} = 1,977\frac{5.00}{2} = 4,942 \text{ pounds per foot}$$

$$f_c = \frac{C}{A} = \frac{4,942}{0.775} = 6,377 \text{ psi}$$

The allowable compressive stress F_c is determined from Eqs. 2.21 through 2.23, depending on the pipe wall slenderness.

$$r = \sqrt{\frac{I}{A}} = \sqrt{\frac{0.0227}{0.775}} = 0.171 \text{ inch}$$

$$\frac{D}{r} = \frac{60.00}{0.171} = 351 > 294 \Rightarrow \text{ use Eq. 2.22}$$

$$F_c = \left[1.212F_y - \frac{E}{3.58 \times 10^8}\left(\frac{D}{r}\right)^2\right]\frac{1}{2} = \left[1.212(33,000) - \frac{29,000,000}{3.58 \times 10^8}351^2\right]\frac{1}{2} = 15,020 \text{ psi}$$

$$\text{Utilization Ratio} = \frac{6,377}{15,020} = 0.42 < 1.00 \ \ \text{O.K.}$$

The calculated compressive stress is less than the allowable stress for this size of pipe, so the pipe is found to be acceptable for the defined conditions and bearing pressure at the surface imposed by the crane.

Other Types of Pipe. The three types of pipe discussed above [plain steel, concrete (both reinforced and nonreinforced), and corrugated steel] are the most common types of buried pipes that a lifting contractor will encounter, particularly on industrial sites. There are other types of pipes with which the contractor and lift planner must be familiar. These include ductile iron pipe (AWWA 2009, DIPRA 2016, U.S. Pipe 2013), PVC and other types of plastic pipe (AWWA 2002, Uni-Bell 2012), and vitrified clay pipe (NCPI 2015).

The design methods for these and other different types of pipe are similar to those discussed in this section. The most important considerations are distinguishing between rigid (like ductile iron) and flexible (like PVC) pipes to assure that the correct pressure calculation model is used and to refer to the appropriate industry design guides and standards, such as those from ASTM International, to assure that the correct section properties and material properties are used in the calculations. Beyond these brief comments, it remains the responsibility of the engineer planning the mobile crane installation to seek out the correct reference material, to understand the behavior of the pipes and related structures, and to apply the design and analysis methods in a manner that is reasonable for the loading developed by the crane.

2.4.3 Strength of Underground Utilities

Evaluation of the strength of underground utilities, such as utility vaults and electrical duct banks, can be significantly more complex than the analysis of buried pipes discussed in the preceding section. Thus, this section will provide only basic guidance and direction.

Precast concrete utility vaults are widely available as manufactured products that are shipped to site and installed in the ground (Fig. 2.13). Vaults may also be specially designed and constructed to meet site-specific requirements. They are covered by a group of industry standards that define loads [ASTM C857-16 (ASTM 2016e)], design [ASTM C478-15a (ASTM 2015b), ASTM C858-10e1 (ASTM 2010)], installation [ASTM C891-11 (ASTM 2011)], and inspection [ASTM C1037-16 (ASTM 2016f)] requirements. The actual concrete design is generally performed in accordance with ACI (2014). Analysis of a utility vault to evaluate its strength with respect to the loads imposed by a mobile crane must be based on these standards.

Figure 2.13 Prefabricated Concrete Utility Vault *(C.R. Barger & Sons, Inc.)*

In the case of a vault that is a manufactured product and assuming that the specific model can identified, the vault manufacturer may be able to assist in the evaluation of the vault under the effects of crane support loading. In the case of a unique design, a structural analysis based on the actual construction may be required. This course of action requires that the appropriate details of the vault construction (dimensions, reinforcing details, material grades, etc.) can be determined reliably.

The design and construction of electrical duct banks is not as well defined. A duct bank is typically a group of pipes, often PVC, that is encased in reinforced concrete (Fig. 2.14). Individual power companies and some cities have their own design requirements for subsurface duct banks, primarily based on electrical safety (i.e. shielding of the cables) rather than load carrying needs. The use of ACI (2014) is reasonable for the analysis of the concrete elements of the duct bank unless the power company, city, or other governing authority defines other requirements. This analysis is also often a soil-structure interaction problem when excessive deflection of the duct bank due to local compression of the soil is a concern.

2.4.4   Bearing Pressure Concentrations

The discussions to this point are based on the assumption of a relatively uniform bearing pressure acting at the surface of the ground. This is not usually the case.

Figure 2.14 Electrical Duct Bank Under Construction *(PJS Electric, Inc.)*

The flexure of the crane mats results in a peak bearing stress concentration directly under the crane outrigger or crawler track, tapering off to either side. This behavior is discussed in great detail in Section 4.2.2. The question here is: How does this peak ground bearing pressure affect the analysis of subsurface pipes and utility structures? The short answer is: Not much. To explain, consider the crane support illustrated in Fig. 2.15.

Fig. 2.15a shows the results of a crane mat analysis based on the balanced design method developed in Chapter 4. The load used in this example is that which produces a ground bearing pressure of 4,000 psf using the effective bearing length calculated following the Chapter 4 method and the assumption of uniform bearing over this effective length. Fig. 2.15b shows the results obtained from the application of a beam on an elastic foundation analysis. The peak bearing pressure is found to be 4,370 psf, which is only 9.25% greater than the average pressure of 4,000 psf. Further, the bearing pressure as illustrated in Fig. 2.15b drops off very rapidly to each side of the crane. (The bearing pressure diagrams in Fig. 2.15 are drawn to scale.) Unless the pipe or utility structure is very close to the surface, the effect at the subsurface plane of the average pressure acting over the full mat area will be more severe than the effect of the small area of peak bearing pressure.

In summary, the use of the average bearing pressure upon which the balanced mat design method developed in Chapter 4 is based for the analysis of subsurface pipes and utility structures generally is a reasonable and acceptable practice. As always, unique situations may call for a different approach.

2.4.5   Retaining Walls and Bulkheads

The need arises occasionally to position a crane near a retaining wall or bulkhead. This can occur when working on an urban site near a building with a basement

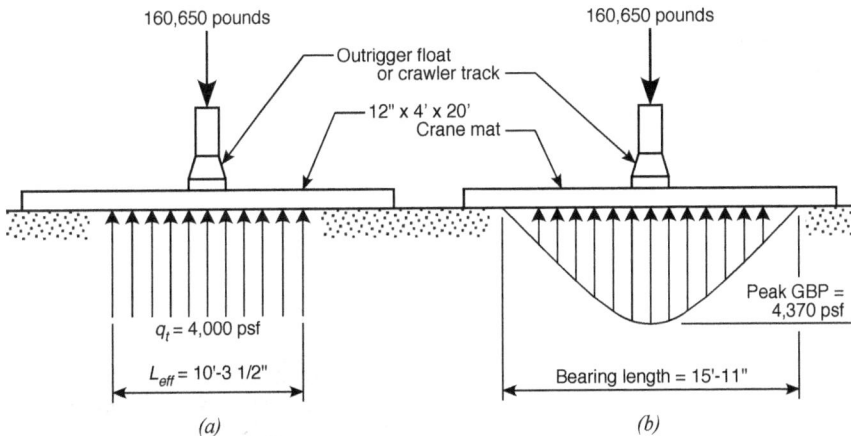

Figure 2.15 Average Ground Bearing Pressure vs. Pressure Concentration

(note that a basement wall is just another type of retaining wall), on a bridge project (near an abutment), or in a port facility. From the point of view of crane support, the engineering issues are similar for all of these situations.

The simplest course of action when faced with any type of retaining structure is to stay back. A substantial vertical load applied to the ground surface near a retaining wall will develop a lateral earth pressure that acts against the surface of the wall. The closer to the wall this vertical load is located, the greater the lateral earth pressure will be. A long-standing rule of thumb (Shapiro and Shapiro 2011, including the three earlier editions) calls for keeping the edge of the matted area that is nearest to the wall back from the base of the wall by at least 1.5 times the height of the wall (Fig. 2.16). This distance can be reduced to be equal to the height of the wall (that is, the dashed line in the figure would be at a slope of 1 : 1) for some stronger, stiffer soils.

If the conditions of the work to be performed do not permit positioning the crane far enough back from the wall or bulkhead, a detailed engineering analysis of the structure must be performed. At a minimum, this analysis must include the following items.

- Calculate the lateral earth pressure acting on the structure due to the weight of the soil. This requires knowledge of certain soil properties, including the density, the angle of internal friction, and the elevation of the water table.
- Calculate the lateral earth pressure acting on the structure due to the support loads from the crane.
- Analyze the bending and shear strength of the structure. This requires having fairly complete details of the structure. In some cases, such as a

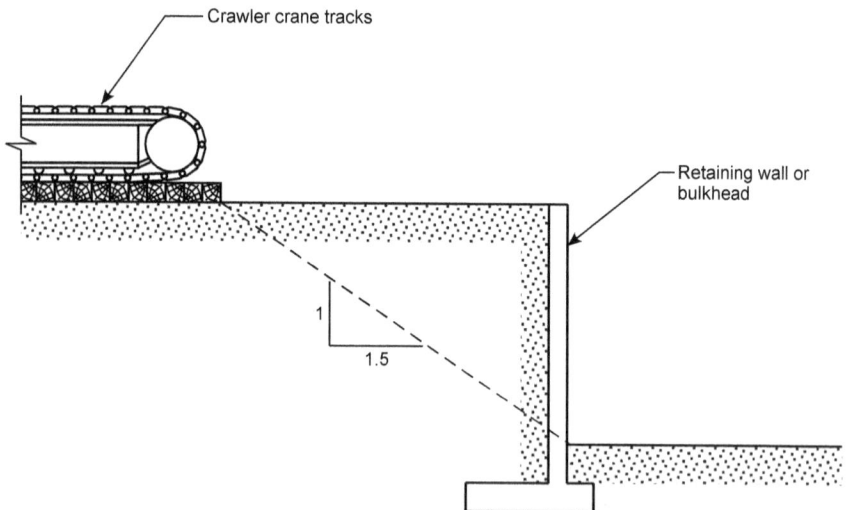

Figure 2.16 Crane Position Near a Retaining Wall

steel sheet pile bulkhead, the properties of the structure can be determined by field measurements. In other cases, such as a reinforced concrete wall, reliable drawings will be needed.

- Analyze the bearing pressures applied to and bearing capacity of the soil under and around the base of the structure.

There are three general ways in which a retaining structure can fail when subjected to excessive lateral earth pressure (Fig. 2.17). These are a structural failure of the wall or its foundation (Fig. 2.17a), a stability failure of the structure in which the entire structure rotates about its base (Fig. 2.17b), and a stability failure in which the entire structure is pushed outward (Fig. 2.17c). We can see that the two stability failures are the result, at least in part, of bearing capacity failures of the soil around the base of the structure.

The methods by which these retaining wall and bulkhead analyses are performed are somewhat complex, but well documented in the existing literature. The equations for the calculation of lateral earth pressures can be found in most foundation engineering textbooks (e.g. Bowles 1996, Das 2016). Specifications for the design of structures include ACI (2014) for reinforced concrete and AISC (2016) for steel structures. Sheet pile bulkhead design is addressed in foundation engineering textbooks, such as Das (2016). There are also specialized references (e.g. Brooks and Nielsen 2013) that address the analysis and design of various types of retaining structures.

In some cases, retaining structures that are not capable of carrying the forces imposed by the crane can be braced temporarily. Diagonal compression struts can be installed between the face of the wall and the ground, provided the means exist to rigidly support the lower ends of these struts. In the case of a trench with walls on both sides, horizontal struts can be used to brace one wall against the other. In all such cases, however, appropriate design engineering of both the bracing struts

Structural failure of wall or its foundation	Rotation of the wall and foundation (overturning)	Outward sliding of the wall and foundation (shear failure at the base)

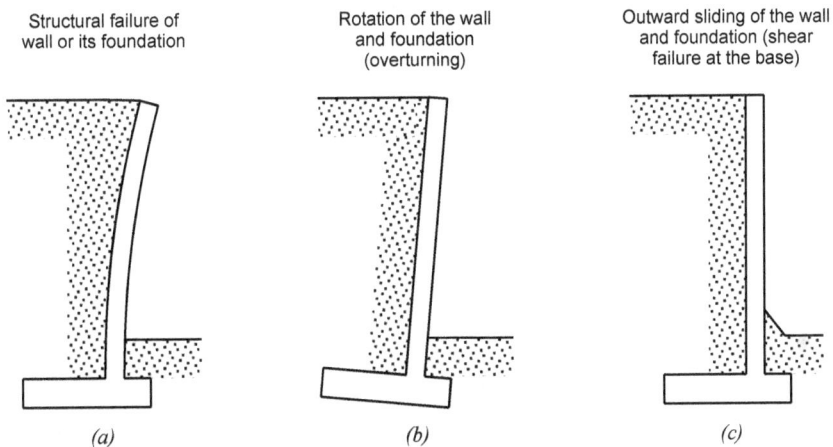

| (a) | (b) | (c) |

Figure 2.17 Retaining Wall Failure Modes

and the retaining structures must be performed to assure that any added braces can carry the resulting loads and that the use of bracing does not compromise the retaining structure in any way.

2.4.6 Embankments

An embankment is a ground surface feature that may be affected by large surface loads. As considered here, an embankment is a relatively abrupt change in the ground surface elevation. While not a structure, the failure mode is such that its treatment with respect to a mobile crane installation is similar to that of a retaining wall or bulkhead, so its inclusion in this section is appropriate. A large surface load near the edge of the embankment can cause a shear failure of the soil, resulting in a collapse of the soil mass under and around the loaded area.

The means of addressing the concerns of working near an embankment is essentially the same as the first suggestion in Section 2.4.5 for working near a retaining wall: Stay back. Fig. 2.18 illustrates a crawler crane on timber mats on the ground surface at the top of an embankment. As with the retaining wall, keeping the end of the loaded area back from the base of the embankment by a distance of at least 1.5 times the height of the embankment is suggested. Again, a 1 : 1 proportion may be acceptable, depending on both the strength of the soil and the shape of the ground surface.

2.5 CRANES SUPPORTED ON STRUCTURES

The previous sections of this chapter address the most common situation of a crane supported on soil. Although structures, such as buried utilities and retaining walls,

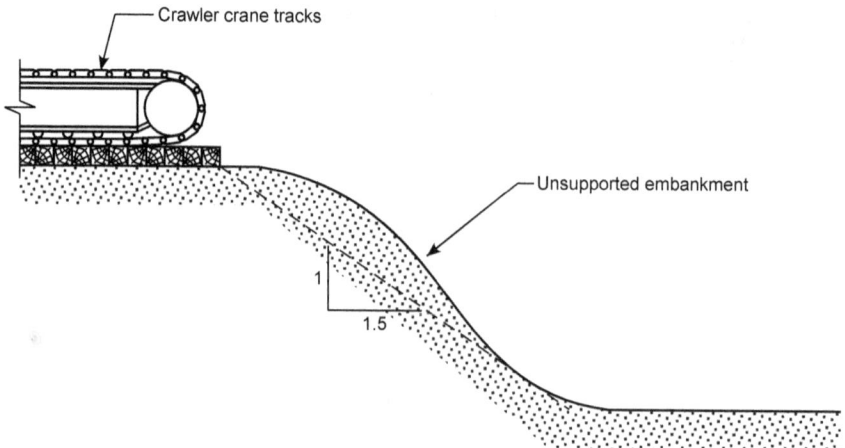

Figure 2.18 Crane Position Near an Embankment

may be affected, the soil is the primary load bearing element. We will now look at the situation in which the crane bears directly on a structure and not on soil. This occurs when a crane is set up on the floor framing of a building, on a bridge, on the deck of a barge, on a large foundation or on paving, such as a road, a parking lot or a building floor that is a slab on grade. Given the very large range of possible structures or components that fall under this heading, only the most general comments will be made here.

The support loads from the crane, whether outrigger reactions or crawler track bearing pressures, are calculated as discussed in Chapter 1. These loads are then applied to the supporting structure, with mats used for load distribution. Section 2.5.1 examines the evaluation of a slab on grade or pavement. Section 2.5.2 examines the distribution of the crane loads when the mats bear on a framed structure.

2.5.1   Pavement and Slabs on Grade

It is often necessary to set up a crane on a paved area. This may be a street, a parking lot, or any other area that is covered by a concrete slab on grade or pavement. In these situations, the strength of the pavement or slab and its ability to distribute the imposed load to the soil below must be evaluated.

Pavements can be broadly divided into two groups: flexible and rigid. Flexible pavement is a layer of material that remains in contact with and distributes loads to the subgrade. This pavement depends on aggregate interlock, particle friction, and cohesion for stability. Bituminous asphalt is the most common type of flexible pavement that a crane user will encounter. Rigid pavement is a structural layer that provides high bending strength and stiffness, allowing it to distribute loads to the subgrade over a comparatively large area. Reinforced concrete is the most common type of rigid pavement.

Regardless of the pavement or slab material, the construction is generally composed of three layers (Fig. 2.19). These are the natural soil, called the subgrade, a compacted subbase (typically sand), and the pavement or slab. The top 6" to 24"

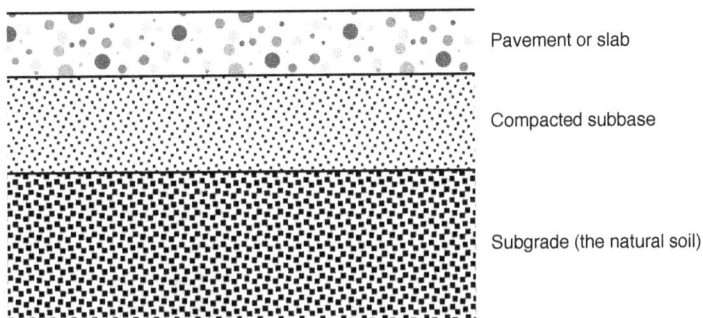

Pavement or slab

Compacted subbase

Subgrade (the natural soil)

Figure 2.19 Construction of Pavement or a Slab on Grade

(150 mm to 600 mm) of the natural soil may be compacted or otherwise treated (e.g. lime stabilization) to improve its load bearing capability prior to placement of the subbase.

When working on new construction, drawings will likely be available on which the details of the pavement or slab will be shown. When working in an existing facility, particularly an older facility, exploration may be necessary to determine the thickness of the pavement, the depth and compaction of the subbase, and the compaction of the subgrade. In the case of a reinforced concrete slab, determination of the size, placement and grade of reinforcing steel, needed to perform an accurate strength analysis, can present a challenge.

Once the necessary dimensions and properties have been determined, the calculation of the load bearing ability of pavement or a slab on grade can be accomplished following established procedures in the published literature. Applicable standards and specifications are listed in Section 2.5.3. Practical considerations for evaluating the load spreading provided by crane mats that bear on pavement or a slab on grade are discussed in Chapter 5.

Use of the standards in Section 2.5.3 requires a cautionary statement here. The concrete slab on grade standard (Farny and Tarr 2008) and the flexible pavement standard (Asphalt Institute 1999) both address the common design cases of wheeled vehicles. Farney and Tarr (2008) also addresses post loads and distributed loads due to stacked materials. The case of a concentrated load bearing on an area in the range of 2 feet (0.6 m) to 4 feet (1.2 m) square, as is the case with an outrigger on a mat, is not explicitly addressed. Thus, some interpretation is necessary in order to apply these standards to a mobile crane installation. However, because the prescribed design methods rely on the use of values pulled from charts and tables, extrapolation to the load magnitudes and bearing area dimensions typical of crane supports may not be straightforward. Alternately, there are specialized software applications available for pavement design that may be useful. As with soil issues in general, this too may be an area in which the lift planner may have to seek out specialized assistance.

2.5.2   Framed Structures

Mobile cranes are often set up on a structure. This may be the floor or roof of a building (Fig. 2.20a), a bridge (Fig. 2.20b), or the deck of a barge or ship (Fig. 2.20c). The engineering of a mobile crane installation for which the support is provided by such a structure may be significantly more complex than that for which the support is provided by soil.

The first challenge is resolution of the crane support loads into reactions applied to the structure. Consider the crawler track illustrated in Fig. 2.21. When bearing on soil or mats, the loading imposed by the track is represented as a pressure diagram that is triangular, trapezoidal, or rectangular. When the crawler is supported on a framed structure, the support provided by that structure will not be

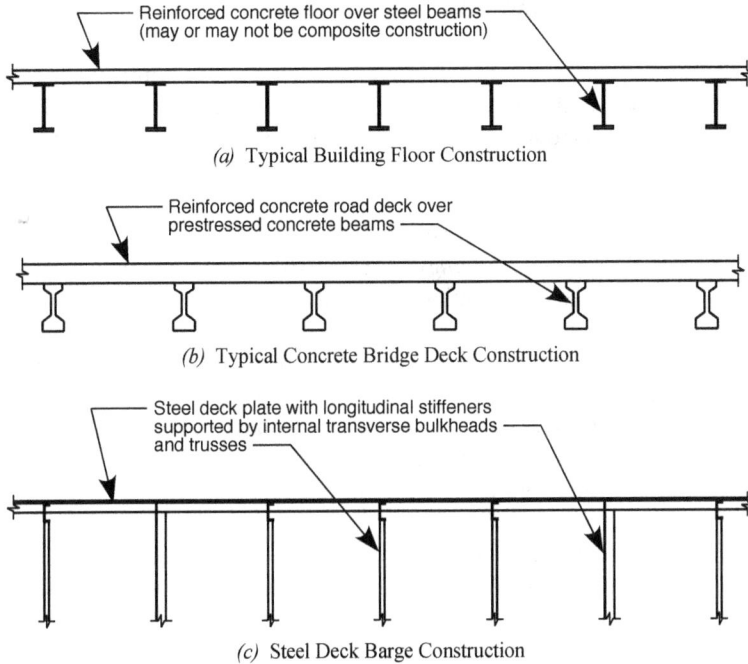

(a) Typical Building Floor Construction

Reinforced concrete floor over steel beams (may or may not be composite construction)

(b) Typical Concrete Bridge Deck Construction

Reinforced concrete road deck over prestressed concrete beams

(c) Steel Deck Barge Construction

Steel deck plate with longitudinal stiffeners supported by internal transverse bulkheads and trusses

Figure 2.20 Common Types of Structures

uniform. Rather, the crawler will bear as concentrated loads at the larger framing members, as illustrated in the figure. The ability of a longitudinal crane mat to spread the load out to the adjacent members not directly loaded by the crawler depends on the relative stiffness of the framing members and the crane mat.

A similar issue exists with support of an outrigger. Unless the floor or deck of the structure or the crane mat is very stiff relative to the vertical stiffness of the framing members of the structure, very little load spreading will occur and the outrigger load may be carried entirely by just one or two members.

The distribution of crane loads into a framed structure is discussed in further detail in Section 5.4.2.

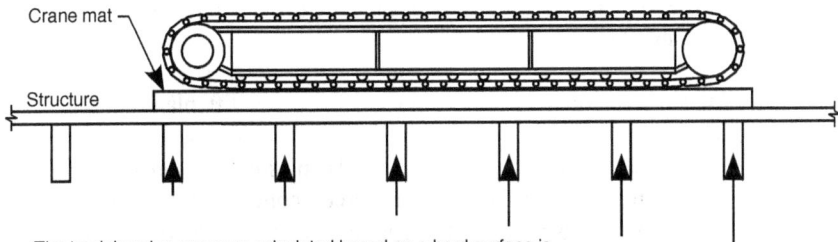

Crane mat

Structure

The track bearing pressure calculated based on a hard surface is resolved into concentrated loads at each hard point of the structure

Figure 2.21 Crawler Track Bearing on a Framed Structure

2.5.3   Applicable Standards

Once the crane loads and other acting loads have been determined, the required evaluation is a structural engineering effort of a magnitude that extends well beyond the scope of this handbook. Listed below, however, are suggested references for the evaluation of various structures.

- Steel building structures are typically analyzed using the AISC *Specification for Structural Steel Buildings* (AISC 2016). Additional guidance on load factors and load combinations for relatively short-term conditions can be found in ASCE (2015).
- Concrete building structures, including many foundation elements, are designed or analyzed using the ACI *Building Code Requirements for Structural Concrete* (ACI 2014). Again, guidance on load factors and load combinations can be found in ASCE (2015).
- Bridge structures are analyzed using the *AASHTO LRFD Bridge Design Specifications* (AASHTO 2017).
- Flexible pavements can be analyzed using the method detailed in MS-1 *Thickness Design – Highways & Streets* (Asphalt Institute 1999). Some interpretation will be necessary since this standard is based on loads being applied to the pavement by wheeled vehicles.
- A concrete slab on grade, which can be pavement or a building floor, can be analyzed using the procedures in *Concrete Floors on Ground* (Farny and Tarr 2008). As with the analysis of flexible pavement, some interpretation and extrapolation will be necessary since this standard is also based on loads applied to the surface by wheeled vehicles, as well as loads from posts.
- Barge structures are proportioned using the *Rules for Building and Classing Steel Barges* (ABS 2017). Part 5, Chapter 3, Section 3 of this document specifically addresses the design of crane barges.
- Most cities throughout the U.S. have local building codes that may have requirements that go beyond those established in the industry standards listed above. The contractor must check with local building officials to assure that any such requirements are met. Also, the owner of the facility where the work is being performed can often provide guidance with respect to applicable local codes.

Some of these structures can be evaluated by many lift planning specialist engineers. The method for analyzing a slab on grade that is developed in Farny and Tarr (2008) is reasonably straightforward and can usually be performed by any engineer who is generally familiar with reinforced concrete design, although as noted, some extrapolation is required to apply this method to loaded areas typical of the footprint of a mobile crane. Likewise, the pavement analysis method in the Asphalt Institute standard is not exceptionally complex, although again requiring interpretation. However, this may not be the case with larger, more complex steel or concrete structures.

In these situations, the contractor may be well advised to work with a structural engineer who is very familiar with building or bridge design and with the codes and standards applicable to the structure at issue. In the case of a barge-mounted crane, the services of a naval architect may be needed to adequately analyze the barge structure.

2.6   APPLICATION SUMMARY

As noted at the beginning of this chapter, geotechnical engineering is a field in which extensive study and experience are required to develop a true expertise. The analysis of subsurface structures, while more straightforward with respect to the calculations employed, is to a great extent based on empirical methods developed from physical testing, as well as on engineering theory. This work likewise requires significant experience to develop a reasonable degree of competence.

The determination of the allowable soil bearing pressure and related soil properties for a given site is addressed in a very general sense in Sections 2.1, 2.2, and 2.3. This material is intended to provide the reader with enough of an understanding of this subject to know what questions to ask. Unless the reader is very knowledgeable in geotechnical engineering, this work is best left to a suitable consultant. The design of special foundation elements for crane support, such as steel piles or drilled shafts, may require assistance from a structural engineer. Again, the brief discussion presented is intended to guide the lift planner, but does not provide thorough coverage of the subject.

Some structures affected by the surface bearing pressures from a mobile crane installation (Section 2.4), such as subsurface pipes, can reasonably be checked by a lift planner who is a civil engineer. This section provides the necessary calculation methods, example problems, and a list of applicable standards to address the most common situations the lift planner is likely to encounter.

Evaluation of a more complex structure on which a crane may be positioned, such as a building floor, a bridge or a barge deck, may be beyond the capabilities of the lift planner (Section 2.5). In these cases, engaging the services of a suitably qualified structural engineer (or naval architect in the case of a barge or ship deck) may be necessary.

In summary, the material covered in this chapter includes calculation methods and tables of data that are appropriate for the engineering of mobile crane installations with respect to ground bearing pressure. However, this is not all there is in this field. The reader is greatly encouraged to become familiar with the various standards referenced in this chapter and to study soils mechanics and foundation engineering textbooks to better understand the principles at work. Calling upon the services of a qualified geotechnical engineer or structural engineer when needed is also strongly encouraged.

2.7   REFERENCES

American Association of State Highway and Transportation Officials (AASHTO) (2017), *AASHTO LRFD Bridge Design Specifications*, 8th ed., Washington, D.C.

American Bureau of Shipping (ABS) (2017), *Rules for Building and Classing Steel Barges*, Houston, TX.

American Concrete Institute (ACI) (2014), 318-14 *Building Code Requirements for Structural Concrete and Commentary*, Farmington Hills, MI.

American Concrete Pipe Association (ACPA) (2011), *Concrete Pipe Design Manual*, 2nd ed., Irving, TX.

American Institute of Steel Construction (AISC) (2016), *Specification for Structural Steel Buildings*, Chicago, IL.

American Iron and Steel Institute (AISI) (1999), *Modern Sewer Design*, 4th ed., Washington, D.C.

American Lifelines Alliance (ALA) (2005), *Guidelines for the Design of Buried Steel Pipe*, Washington, D.C.

American Society of Civil Engineers (ASCE) (2015), SEI/ASCE 37-14 *Design Loads on Structures During Construction*, Reston, VA.

American Water Works Association (AWWA) (2002), M23 *PVC Pipe – Design and Installation*, 2nd ed., Denver, CO.

American Water Works Association (AWWA) (2017), M11 *Steel Pipe – A Guide for Design and Installation*, 5th ed., Denver, CO.

American Water Works Association (AWWA) (2009), M41 *Ductile-Iron Pipe and Fittings*, 3rd ed., Denver, CO.

Asphalt Institute (1999), MS-1 *Thickness Design – Highways & Streets*, 9th ed., Lexington, KY.

ASTM International (2010), C858-10e1 *Standard Specification for Underground Precast Concrete Utility Structures*, West Conshohocken, PA.

ASTM International (2011), C891-11[A] *Standard Practice for Installation of Underground Precast Concrete Utility Structures*, West Conshohocken, PA.

ASTM International (2015a), C14-15a[A] *Standard Specification for Nonreinforced Concrete Sewer, Storm Drain, and Culvert Pipe*, West Conshohocken, PA.

ASTM International (2015b), C478-15a[A] *Standard Specification for Circular Precast Reinforced Concrete Manhole Sections*, West Conshohocken, PA.

ASTM International (2015c), C985-15[A] *Standard Specification for Nonreinforced Concrete Specified Strength Culvert, Storm Drain, and Sewer Pipe*, West Conshohocken, PA.

ASTM International (2016a), C76-16[A] *Standard Specification for Reinforced Concrete Culvert, Storm Drain, and Sewer Pipe*, West Conshohocken, PA.

ASTM International (2016b), C506-16b[A] *Standard Specification for Reinforced Concrete Arch Culvert, Storm Drain, and Sewer Pipe*, West Conshohocken, PA.

[A] These standards use U.S. customary units. The metric (SI units) versions are indicated with an "M" appended to the designation, e.g. C891M-11.

ASTM International (2016c), C507-16[A] *Standard Specification for Reinforced Concrete Elliptical Culvert, Storm Drain, and Sewer Pipe*, West Conshohocken, PA.

ASTM International (2016d), C655-16[A] *Standard Specification for Reinforced Concrete D-Load Culvert, Storm Drain, and Sewer Pipe*, West Conshohocken, PA.

ASTM International (2016e), C857-16 *Standard Practice for Minimum Structural Design Loading for Underground Precast Concrete Utility Structures*, West Conshohocken, PA.

ASTM International (2016f), C1037-16 *Standard Practice for Inspection of Underground Precast Concrete Utility Structures*, West Conshohocken, PA.

ASTM International (2016g), D1196-12(2016)[A] *Standard Test Method for Nonrepetitive Static Plate Load Tests of Soils and Flexible Pavement Components, for Use in Evaluation and Design of Airport and Highway Pavements*, West Conshohocken, PA.

Bowles, J.E. (1996), *Foundation Analysis and Design*, 5th ed., The McGraw-Hill Companies, Inc., New York, NY.

Brooks, H., and Nielsen, J.P. (2013), *Basics of Retaining Wall Design*, 10th ed., HBA Publications, Inc., Corona del Mar, CA.

Construction Industry Research and Information Association (CIRIA) (2003), *Crane Stability on Site*, 2nd ed., London, U.K.

Das, B.M. (2016), *Principles of Foundation Engineering*, 8th ed., Cengage Learning, Stamford, CT.

Ductile Iron Pipe Research Association (DIPRA) (2016), *Design of Ductile Iron Pipe*, Golden, CO.

Farny, J.A., and Tarr, S.M. (2008), EB075 *Concrete Floors on Ground*, 4th ed., Portland Cement Association, Skokie, IL.

Moser, A.P., and Folkman, S. (2008), *Buried Pipe Design*, 3rd ed., The McGraw-Hill Companies, New York, NY.

National Clay Pipe Institute (NCPI) (2015), *Vitrified Clay Pipe Engineering Manual*, Elkhorn, WI.

Naval Facilities Engineering Command (NAVFAC) (1986a), Design Manual DM-7.01 *Soil Mechanics*, U.S. Government Printing Office, Washington, D.C.

Naval Facilities Engineering Command (NAVFAC) (1986b), Design Manual DM-7.02 *Foundations and Earth Structures*, U.S. Government Printing Office, Washington, D.C.

Shapiro, L.K., and Shapiro, J.P. (2011), *Cranes and Derricks*, 4th ed., The McGraw-Hill Companies, Inc., New York, NY.

Shastri, A., and Sanchez, M. (2012), "Mechanical Modeling of Frozen Soils Incorporating the Effect of Cryogenic Suction and Temperature," *Unsaturated Soils: Research and Applications*, Mancuso, C., Jommi, C., and D'Onza, F., eds., Springer-Verlag Berlin Heidelberg.

Uni-Bell PVC Pipe Association (2012), *Handbook of PVC Pipe Design and Construction*, 5th ed., Dallas, TX.

U.S. Pipe and Foundry Company (2013), *Ductile Iron Pipe Design*, Birmingham, AL.

Yong, R. (1963), "Research on Fundamental Properties and Characteristics of Frozen Soils," pp. 84-108, *Proceedings of the First Canadian Conference on Permafrost*, Technical Memorandum No. 76, National Research Council of Canada, Ottawa.

3　Crane Mat Strength and Stiffness

Safe use of a mobile crane requires that the crane be supported in such a manner that the strength of the supporting surface is not exceeded and that the crane will remain within the required degree of level during its operation. When the crane is set up on soil, crane mats are often required to spread the loads imposed by the crane over a sufficiently large bearing area. In some cases, crane mats used with relatively small cranes may be selected based on rules of thumb that have developed over the years (see Section 5.6). Mats used under large cranes generally must be designed using appropriate engineering calculations to assure that the strength and stiffness required to support the crane adequately are provided.

One aspect of crane mat design that bears investigation is the load spreading performance that is provided by the timber mats commonly in use on construction work sites. Timber crane mats are not precision-manufactured products and there are no established standards that guide their construction. Rather, a crane mat is assembled from rough-sawn timbers of various species using steel tie rods of various diameters.

The first two sections of this chapter present the results of studies that were undertaken by the author to quantify the dimensions, configurations, conditions, and wood species of timber crane mats as used in construction throughout the United States by examining a broad sample of mats being used by contractors and by surveying crane mat manufacturers. Subsequent sections then examine wood industry standards combined with the study results to determine appropriate strength and stiffness properties that may be used for the analysis and design of crane mats.

Most of this chapter is dedicated to an examination of the strength properties of timber crane mats, as this is the type of mat most commonly used with mobile cranes. The chapter concludes with brief discussions of fabricated steel crane mats and crane mats made of synthetic materials, which are typically proprietary products.

3.1　TIMBER MAT DIMENSIONS AND CONDITION SURVEY

The calculation of the strength of a crane mat requires an understanding of its physical properties. These properties include the dimensions of the timbers, the size, number and spacing of the steel tie rods, and the expected condition of the wood. In order to quantify these properties, the author conducted a survey of the dimensions and condition of timber crane mats in service. This nationwide survey was performed between November 2009 and December 2012.

3.1.1 Methodology

The author contacted construction contractors and engineers around the United States and requested the taking of measurements of timber crane mats. The information was to be entered on a standard form (Fig. 3.1), with the timber depths reported to within 1/16" and the overall mat width and tie rod spacings reported to within 1/8" (all measurements were reported in USCU). The respondents were also asked to provide comments about the condition of the mats, their age (if known), and to take end and side view photographs of each mat so measured.

Location (city, state): _____

Mat Size: _____

Timber Dimensions:

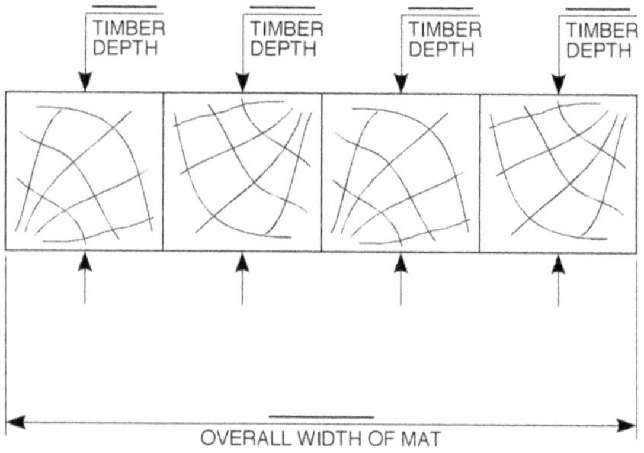

Total Number of Tie Rods: _____   Tie Rod Diameter: _____

Tie Rod Spacing, maximum: _____   Tie Rod Spacing, minimum: _____

Comments on Mat Condition:

_____

Figure 3.1 Crane Mat Dimension Survey Form

The locations from which data were obtained are illustrated on the map in Fig. 3.2. A total of 165 mats (642 timbers) were measured in this study. Most of the mats (94%) were of a nominal thickness of 12" (300 mm). The remaining specimens were either 8" (200 mm) or 10" (250 mm) thick.

All of the measurements collected in this survey were reported in feet and inches. Dimensions shown here in SI units are all converted values.

3.1.2   Survey Results

Of the 608 timbers in the mats with a nominal thickness of 12" (300 mm), the maximum depth was found to be 13.125" (333 mm), the minimum was 10.000" (254 mm), and the average depth is calculated to be 11.741" (298 mm). The standard deviation is 0.403" (10.2 mm) and the coefficient of variation is 0.034. We can show the depth data for this group as the ratio of the measured depth to the nominal depth of 12". In this form, the average ratio is 0.978. The depth distribution is illustrated as a histogram in Fig. 3.3.

The measured widths of the mats were also found to be less than the nominal widths. Using the overall mat width data, the average timber width is 11.733" (298 mm), thus giving a ratio of measured width to nominal width of 0.978 and a coefficient of variation of 0.034, the same as that of the depth measurements. (As seen in Fig. 3.1, only the overall width of the mat was measured; the widths of the individual timbers were not measured.)

Wherever the timbers were not perfectly square, the mat manufacturers generally oriented them in the mat such that the greater dimension was the depth.

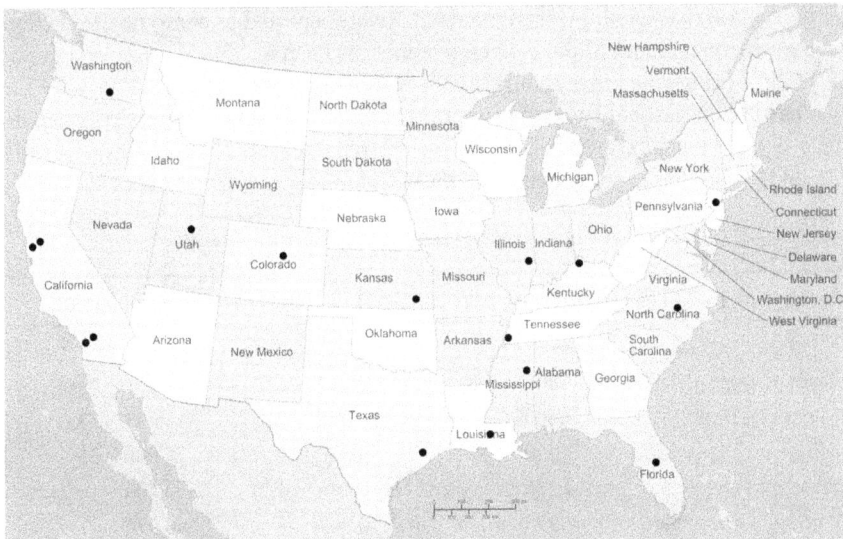

Figure 3.2 Dimension Survey Response Locations

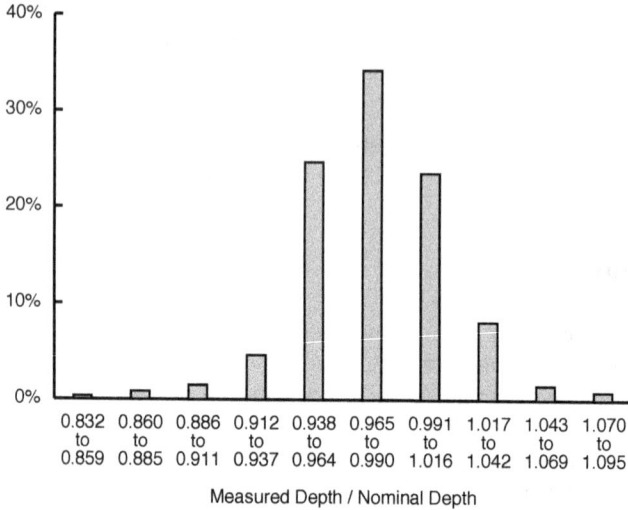

Figure 3.3 Timber Depth Ratio Distribution Histogram

The samples of mats with 8" (200 mm) and 10" (250 mm) timbers are not great enough to allow independent evaluations. In general, however, the measurements from those groups exhibit average values and scatters that are reasonably consistent with those reported for the 12" (300 mm) timbers. A summary of the depth measurements of all timbers is shown in Table 3.1.

Considering the deviations of both timber depth and timber width, the section properties for bending performance can also be examined. Since the widths of individual timbers were not measured, we will use average timber widths based on the overall mat width measurements that were reported to compute the section properties. A summary of the moments of inertia (I_x) in in.^{4} (mm^4), section moduli (S_x) in in.^{3} (mm^3), and area in in.^{2} (mm^2) is shown in Table 3.2.

The timbers are held together to form a mat by means of threaded steel tie rods. Tie rod diameters reported in the survey ranged from 3/4" (19 mm) to 1 3/8" (35 mm), with 1" (25 mm) being the most common size. Tie rod spacing varied

TABLE 3.1 Summary of Timber Dimension Survey Results - Depth

Mat Size	8" (200 mm)	10" (250 mm)	12" (300 mm)
Number of Specimens	28	6	608
Average Depth	7.804" (198)	9.667" (246)	11.741" (298)
Maximum Depth	8.500" (216)	10.000" (254)	13.125" (333)
Minimum Depth	7.250" (184)	9.500" (241)	10.000" (254)
Standard Deviation	0.255" (6.5)	0.186" (4.7)	0.403" (10.2)
Average / Nominal	0.975	0.967	0.978
Coefficient of Variation	0.033	0.019	0.034

TABLE 3.2 Summary of Timber Dimension Survey Results - Section Properties

Mat Size	8" x 48" (200 x 1,200)	10" x 48" (250 x 1,200)	12" x 48" (300 x 1,200)
Average I_x	1,912.6 (796,070,000)	3,552.7 (1,478,700,000)	6,379.5 (2,655,300,000)
Maximum I_x	2,103.0 (875,330,000)	3,733.0 (1,553,800,000)	7,998.1 (3,329,100,000)
Minimum I_x	1,607.0 (668,880,000)	3,394.0 (1,412,700,000)	4,988.8 (2,076,500,000)
Standard Deviation	158.5 (65,970,000)	139.2 (57,960,000)	540.0 (224,800,000)
Average / Nominal	0.934	0.888	0.923
Coefficient of Variation	0.083	0.039	0.085
Average S_x	477.5 (7,825,000)	728.5 (11,940,000)	1,061.9 (17,400,000)
Maximum S_x	512.0 (8,390,000)	746.7 (12,240,000)	1,226.6 (20,100,000)
Minimum S_x	421.4 (6,906,000)	714.5 (11,710,000)	886.9 (14,530,000)
Standard Deviation	28.3 (463,500)	13.5 (220,700)	64.6 (1,058,000)
Average / Nominal	0.933	0.911	0.922
Coefficient of Variation	0.059	0.018	0.061
Average Area	375.6 (242,300)	455.9 (294,100)	553.3 (356,900)
Maximum Area	392.0 (252,900)	459.2 (296,300)	603.3 (389,200)
Minimum Area	339.8 (219,200)	451.3 (291,200)	495.1 (319,400)
Standard Deviation	16.2 (10,450)	3.4 (2,160)	20.7 (13,380)
Average / Nominal	0.978	0.950	0.961
Coefficient of Variation	0.043	0.007	0.037

greatly, even within individual mats. The spacing ranged from 7.3% of the overall mat length to 71.5% of the mat length. The average is 24.4% of the mat length and 82% of the mats had tie rod spacings that were in the range of 15% to 30% of the mat length.

The conditions of the crane mats were reported both by written descriptions and through photographs. Mats were most commonly described as being in good condition with local damage to the edges. This is understandable, as the edges are regularly subjected to abuse as crawler cranes move on and off the mats.

The condition of the mats seen in Fig. 3.4 is typical of the many photographs received from the survey. Many vendors or owners paint the ends of the timbers to seal the ends and for identification. Splits, as seen in the middle mat in Fig. 3.4*a*, are fairly common. Other characteristics are discussed and illustrated in Section 3.3.3. In all cases, the mats showed signs of weathering, to be expected since mats are normally stored outdoors without protection from the elements.

It is understood that timber mats will deteriorate over time, both from use and from exposure. This creates a demand on the mat users to regularly inspect their mats and to discard mats that show significant degradation. However, there are presently no standards that establish what constitutes "significant degradation." It remains the responsibility of the mat owners and users to exercise good judgment in this regard.

3.2 TIMBER SPECIES SURVEY

The strength and stiffness of a timber is a function of the species of wood, as well as of its size and its condition. A survey of fifteen crane mat vendors throughout the United States was conducted in late 2013 and early 2014 to determine the species of the timbers used in the production of crane mats. This survey was extended in 2015 and 2017 to cover five additional vendors for a total of 22 locations.

Crane mat timbers are most commonly identified as mixed hardwoods, with about three-quarters of the vendors identifying the wood used as such. ("Mixed hardwoods" as a commercial species group was used in wood standards up through the mid-1980s. This group is discussed further in Section 3.5.1.) The second most common species identified was oak, with one vendor specifically indicating red and white oak. Almost half of the vendors surveyed offer mats made of oak timbers. Hickory and hemlock timbers were also identified by two vendors. Douglas fir is used for crane mat construction by six vendors. Two vendors identified the timbers used as mora and one vendor uses azobe timbers. Some crane mat vendors outside of the U.S. also use mora and azobe timbers, as well as other tropical

(a)   *(b)*

Figure 3.4 Typical Condition of Crane Mats in Service *(David Duerr, P.E.)*

hardwoods. Given the stated geographic limit of the present survey, only the two domestic vendors are included here.

The results of the timber species survey are presented in Table 3.3. The figures in the table indicate the number of crane mat production locations identified by state that stated the use of the various wood species or species groups. The numbers in Table 3.3 don't add up to 22 since most mat vendors use more than one species of wood in their mats. The species are organized in the table by group, with the hardwoods (hickory, mixed hardwoods, and oak) first, the softwoods (Douglas fir, hemlock, pine, and other softwoods) second, and the two tropical hardwoods (azobe and mora) last.

We can surmise from this information that the specific species of the timbers in a crane mat are almost impossible to identify, even in a new mat (unless, of course, the mat vendor can supply this information). This observation will figure in our evaluation of the strength of a crane mat as we proceed through this chapter.

On a few occasions over the years, the author has come across mat vendors that use species of lower strength than those listed in Table 3.3. These species include cottonwood, hackberry, and sweet gum. In addition to having lower strength characteristics than the more common species, some users have reported that these species degrade more quickly in the field, thus leading to a shorter useful life.

TABLE 3.3 Summary of Timber Species Survey Results

Vendor Location	Hickory	Mixed Hardwoods	Oak	Douglas Fir	Hemlock	Pine	Other Softwoods	Azobe	Mora
Alabama								1	
California (northern)		1		1					
Florida	1	2	2						1
Georgia		1	1						
Illinois		2	2	1					
Louisiana		2	1						
Minnesota		1	1	1					
Mississippi		1							
New Jersey		1		1		1			
North Carolina		1	2						
Oregon		3		3	1	1			
Texas		1	2						1
Washington		1					1		
Percent of Locations using the Species	5	77	50	32	5	9	5	5	9

The detailed discussions in Sections 3.3 and 3.5 are based on mats made of the timber species listed in Table 3.3 other than azobe and mora, based on their dominance in the U.S. Discussions of the strength characteristics of other species and how mats of these species should be treated are given in Sections 3.5.2 and 3.5.6. The challenge on the part of the user is to recognize the use of these different species and to adjust the mat design calculations accordingly.

3.3   STANDARDS FOR TIMBER PROPERTIES

Wood is a natural material, not a manufactured material like steel or aluminum. Consequently, the physical properties of a piece of wood cannot be established through the manufacturing process. Rather, the properties of a timber must be determined by inspecting the piece to evaluate its characteristics relative to characteristics known to affect strength. This inspection process culminates in the assigning of a grade to each piece of wood. That grade correlates to strength properties established in various design standards.

The methods of inspecting and grading structural timbers are defined in a number of standards published by ASTM and by the lumber industry. This section presents a brief discussion of the standards that are applicable to the types and grades of timbers commonly used in the construction of crane mats.

3.3.1   ASTM Standards

ASTM Volume 04.10 contains over 90 standards applicable to the testing, inspection, and grading of wood products. Of these, only three are directly applicable to the types and species of timbers used in the manufacture of crane mats. These are D9-12 (ASTM 2012), D245-06(2011) (ASTM 2011), and D2555-17a (ASTM 2017). The editions shown here and listed in the References section at the end of this chapter are those that are current as of this writing. Like most standards, they are revised from time to time.

ASTM D9-12 *Standard Terminology Relating to Wood and Wood-Based Products* is, as the title indicates, a list of definitions of the terms used in the other wood standards. Of particular value is the appendix that illustrates the means of determining the size of a knot.

ASTM D245-06 (reapproved 2011) *Standard Practice for Establishing Structural Grades and Related Allowable Properties for Visually Graded Lumber* defines a method of determining the reduction in strength properties of a piece of wood based on the presence of defects in the piece. This is a fundamentally important process. In summary, the greatest strength of a particular species of wood is found in a clear specimen; that is, a piece of wood with no defects. An example of such a specimen is the white oak dowel pictured in Fig 3.5. Defects such as sloped grain, knots, and the like decrease the strength of the piece. D245-06 provides methods of accounting for the strength-reducing effects of those defects. We will examine the specifics of this process in Section 3.5.

ASTM D2555-17a *Standard Practice for Establishing Clear Wood Strength Values* defines test methods and procedures for determining the strength properties of clear wood specimens. The most useful content relative to this discussion are the tables of clear wood strength values for dozens of species of wood. The species covered in this standard include all of the hardwood and softwood species shown in Table 3.3 except azobe and mora. For reference, comparable values for azobe and mora can be found in the *Wood Handbook* (FPL 2010). The clear wood strength values from D2555-17a form the input for the calculations defined in ASTM D245-06.

When code-specified strength values are used for design, these ASTM standards serve to provide an understanding of the basis for those values. When faced with a unique design application, such as evaluating the strength of a crane mat made of tropical hardwood timbers (e.g. azobe or mora), these standards can be used to develop strength values that are appropriate for that application, while still being consistent with established design practices.

3.3.2   Industry Standards and Guides

The types of lumber of interest here are visually graded at the mills as the first step in the determination of their strength properties. The lumber grading industry in the United States and Canada operates under the umbrella of the American Lumber Standard Committee, Inc. (ALSC), a non-profit company based in Germantown, Maryland. ALSC oversees the work of seven industry organizations that develop and maintain sets of lumber grading rules. These organizations are:

- National Lumber Grades Authority (NLGA)
- Northeastern Lumber Manufacturers Association (NeLMA)
- Northern Softwood Lumber Bureau (NSLB)

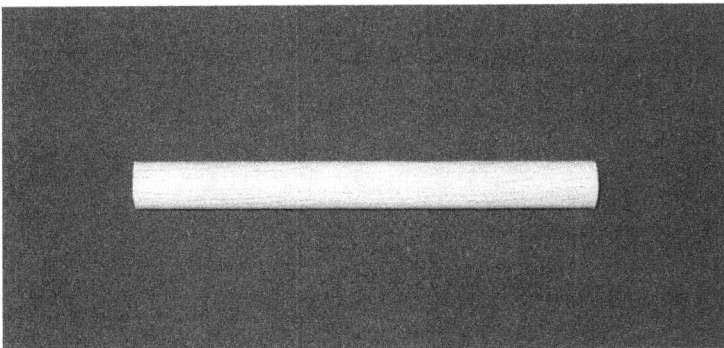

Figure 3.5 White Oak Dowel *(David Duerr, P.E.)*

- Redwood Inspection Service (RIS)
- Southern Pine Inspection Bureau (SPIB)
- West Coast Lumber Inspection Bureau (WCLIB)
- Western Wood Products Association (WWPA)

The first organization on this list is Canadian. The other six are based in the United States.

Of the inspection and grading publications from these seven organizations, only NLGA (2014), NeLMA (2013), and WCLIB (2015) are applicable to the size classification (posts and timbers) and species (various structural hardwoods and softwoods) of timbers that are commonly used in the construction of crane mats. The grading provisions defined in these three publications are essentially identical for the three stress grades of posts and timbers, identified as Select Structural, No. 1, and No. 2. The grading requirements for these three grades of posts and timbers are assembled in Table 3.4.

Some of the characteristics that form the basis for grading do not affect strength. Rather, they affect appearance, which may be important for timbers used as exposed structural members in a building. Since appearance is not a concern for crane mat performance, these characteristics may be ignored. Of the characteristics listed in Table 3.4, only checks, knots, shake, slope of grain, splits, and wane are considered in the establishment of allowable strength and stiffness properties of timbers.

3.3.3   Application of the Timber Standards - Defects

The standards discussed in Sections 3.3.1 and 3.3.2 combine to give us a means of visually inspecting and grading the timbers that are used in the construction of crane mats and determining the expected strength and stiffness of the various grades. In this section, we will discuss the significant growth characteristics that affect timber performance.

The ultimate goal of this process is the determination of the strength ratio of a timber, which is the ratio of its actual strength considering the defects present to the strength that the piece would have if it was free of defects. In this process, a strength ratio for each type of stress (i.e., bending, horizontal shear, etc.) is determined for each type of defect present and then the lowest value for each type of stress is used in the determination of each allowable stress. This is where we combine the allowable defects defined by the grading rules with the calculation method established by ASTM D245-06 (ASTM 2011).

Checks. A check is a separation of the wood along the fiber direction (the long dimension in a timber) that usually extends across the annual growth rings. Checks only affect horizontal shear strength in bending members. The strength loss is assumed to be based on a loss of cross section due to the size of the check. ASTM

D245-06 prescribes a strength ratio of 50% for all sizes of checks for all sizes of lumber. This is the maximum effect that a check can have on the horizontal shear strength of a bending member. We can see that the shear strength ratio for checks is independent of the grade of the timber.

Knots. A knot is a portion of a limb of the tree that has been surrounded by subsequent growth of the wood. The strength loss in bending due to the presence of a knot is based on the reduction of the section modulus of the timber due to the loss of the area occupied by the knot. The effect is greatest when the knot is on the tension side of a bending member.

The greatest strength loss occurs with an edge knot, as shown in Fig. 3.6. In this case, the bending strength, and therefore the strength ratio, is based on the net section modulus below the knot. The knot seen dimensioned in Fig. 3.6 is at the high end of the knot sizes observed in the timber dimension study. (The boundary of this knot is defined by cracks around the edge of the knot. These cracks are not easily seen in this half-tone printed version of the photograph.) The timber in Fig. 3.6 is (nominally) 12" deep, so the depth below the knot is 9.75". The section modulus below the knot is 190 in.^{3} and the section modulus of the gross section is 288 in.^{3}. This equates to a strength ratio of 190 / 288 x 100% = 66.0%.

The strength ratios for posts and timbers, which are based on the knot size and the width of the wide face, are calculated using the formulas given in Appendix X1.3 of ASTM D245-06.

Shake. Shake is a longitudinal separation of the wood. Shake is treated identically to checks by D245-06 in that the strength reduction applies only to horizontal shear. The strength ratio for shake is to be taken as 50% for all sizes and grades of timbers.

Figure 3.6 Edge Knot in a Crane Mat Timber *(David Duerr, P.E.)*

TABLE 3.4 Visual Grading Requirements for Posts and Timbers

Characteristic	Select Structural	No. 1	No. 2
Checks	Seasoning checks, single or opposite each other with a sum total equal to 1/2 the thickness of the piece.	Seasoning checks, single or opposite each other with a sum total equal to 1/2 the thickness.	Seasoning checks.
Knots	Sound, tight, and well spaced. Knot size limitations are permitted in the following sizes or their equivalent displacement.	Sound, tight, and well spaced. Knot size limitations are permitted in the following sizes or their equivalent displacement.	Sound, not firmly fixed, or holes, well spaced. Knot size limitations are permitted in the following sizes or their equivalent displacement. Unsound knots are limited to 1/2 the size of other knots.
	Wide Face Width — Knot Size 5" (127) — 1" (25) 6" (152) — 1 1/4" (32) 8" (203) — 1 5/8" (41) 10" (254) — 2" (51) 12" (305) — 2 3/8" (60) 14" (356) — 2 1/2" (64) 16" (406) — 2 3/4" (70) 18" (457) — 3" (76)	Wide Face Width — Knot Size 5" (127) — 1 1/2" (38) 6" (152) — 1 7/8" (48) 8" (203) — 2 1/2" (64) 10" (254) — 3 1/8" (79) 12" (305) — 3 3/4" (95) 14" (356) — 4" (102) 16" (406) — 4 1/4" (108) 18" (457) — 4 1/2" (114)	Wide Face Width — Knot Size 5" (127) — 2 1/2" (64) 6" (152) — 3" (76) 8" (203) — 3 3/4" (95) 10" (254) — 5" (127) 12" (305) — 6" (152) 14" (356) — 6 1/2" (165) 16" (406) — 7" (178) 18" (457) — 7 1/2" (191)
Pin Holes	Limited.	Limited.	Not limited.
Pitch Streaks	Not limited.	Not limited.	Not limited.
Pockets	Medium pitch pockets.	Pitch pockets.	Pitch or bark pockets.

TABLE 3.4 Visual Grading Requirements for Posts and Timbers (con't.)

Characteristic	Select Structural	No. 1	No. 2
Shake	1/3 the thickness on end.	1/3 the thickness on end.	1/2 length, 1/2 thickness. If through at ends, limited as splits.
Skips	Occasional skips 1/16" (1.6) deep, 2' (610) in length.	Occasional skips 1/8" (3.2) deep, 2' (610) in length.	1/8" (3.2) deep, 2' (610) in length, or 1/16" (1.6) skip full length.
Slope of Grain	1 in 12.	1 in 10.	1 in 6.
Splits	Splits equal in length to 3/4 the thickness of the piece or equivalent of end checks.	Splits equal in length to the width of the piece or equivalent of end checks.	Medium or equivalent of end checks.
Stain	Stained sapwood. Firm heart stain, 10% of width or equivalent.	Stained sapwood. Firm stained heartwood.	Stained wood.
Torn Grain	Heavy.	Heavy.	Not limited.
Unsound Wood			Small spots of unsound wood, well scattered, 1/6 the face width.
Wane	1/8 of any face, or equivalent slightly more for a short distance.	1/4 of any face, or equivalent slightly more for a short distance.	1/3 of any face, or equivalent slightly more for a short distance.
White Speck			Firm white specks, 1/3 width or equivalent.

Slope of Grain. The slope of the grain is, as the name implies, the angle of the grain relative to the sides of the piece of lumber. The bending strength ratios given in D245-06 for the three grades of posts and timbers discussed here and listed in Table 3.4 are 40% for a
slope of grain equal to 1 in 6, 61% for a slope of 1 in 10, and 69% for a slope of 1 in 12. Fig. 3.7 graphically illustrates grain slopes that correspond to these three values. Fig. 3.8 is a side view of two crane mats in which can be seen the grain of the timbers. Comparing the grain seen in Fig. 3.8 to the slope of the lines in the drawings of Fig. 3.7, we can see that the slope of grain is fairly flat, certainly not greater than the 1 in 12 slope permitted for Select Structural timbers. The grain pictured in Fig. 3.8 is typical for the mats inspected for the survey.

The strength ratios given in D245-06 for the various slopes of grain are based on experimental research. Short local deviations of slope of grain, such as those that occur around knots, are not considered in determining the slope of grain of a piece of lumber for the purpose of determining the strength ratio. Slope of grain only affects timber strength in bending, tension parallel to the grain, and compression parallel to the grain. Of these, only bending is applicable to the evaluation of the strength of a crane mat. There is no strength ratio for horizontal shear associated with slope of grain.

Splits. A split is a separation of the wood parallel to the fiber direction caused by the tearing apart of the wood cells. As with shake, splits are also treated identically to checks by D245-06 in that the strength reduction applies only to horizontal shear and the strength ratio is to be taken as 50% for all sizes and grades of timbers. (Splits can be seen in some of the timbers pictured in Fig. 3.4a.)

Wane. Wane is the presence of bark or the lack of wood along an edge or corner of a piece of lumber. Wane can be seen along the lower edge of the middle mat in

(a) Select Structural - 1 : 12 Grain Slope

(b) No. 1 - 1 : 10 Grain Slope

(c) No. 2 - 1 : 6 Grain Slope

Figure 3.7 Graphical Representation of Grain Slopes

Figure 3.8 Side View of Crane Mats *(David Duerr, P.E.)*

Fig. 3.8. Since this loss of section obviously corresponds to a reduction in both bending and shear strength, one may reasonably ask why wane is not among the defects to which D245-06 assigns strength ratios. The reason is simple: It isn't significant. This is demonstrated below by means of an example using proportions applicable to a crane mat timber.

EXAMPLE 3-1

Consider the square timber illustrated in Fig. 3.9. The timber is identified as 12" x 12" (305 mm x 305 mm) and each corner has lost a 1 1/2" x 1 1/2" (38 mm x 38 mm) chamfer due to wane. The total loss on each face is thus 2 x 1 1/2" = 3" (76 mm), which is 3" ÷ 12" = 1/4 of the width of the face. This is the limit for wane for No. 1 posts and timbers (see Table 3.4). Calculating the bending strength (based on the section modulus) and the shear strength (based on the cross sectional area) of this timber, taking into account the section loss due to wane, we find the following:

Full timber S_x = 288.00 in.^{3} (4,720,000 mm^3)
Reduced timber S_x = 265.22 in.^{3} (4,350,000 mm^3)
Bending Strength Ratio = 265.22 / 288.00 x 100% = 92.1%

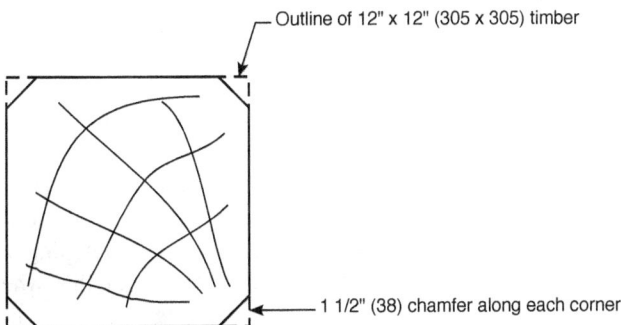

Outline of 12" x 12" (305 x 305) timber

1 1/2" (38) chamfer along each corner

Figure 3.9 The Effect of Wane

Full timber shear area = 144.00 in.^{2} (92,900 mm^2)
Reduced timber shear area = 139.50 in.^{2} (90,000 mm^2)
Shear Strength Ratio = 139.50 / 144.00 x 100% = 96.9%

The strength ratios seen in Example 3-1 are significantly greater than the strength ratios for other defects, such as knots for bending and checks for shear, thus showing us that the strength loss in a timber due to wane is much less than that due to other defects that are permitted by the grading rules. Since the strength ratios are not additive (only the lowest value applies), the strength loss due to wane can be safely ignored in determining allowable stresses. For reference, the bending and shear strength ratios for the amount of wane permitted for No. 2 posts and timbers are 86.7% and 94.4%, respectively. However, strength loss due to wane can come into play if the need arises to evaluate the strength of a mat in which the timbers show significantly increased wane as a result of damage acquired from use.

Quality Factor for Modulus of Elasticity. The modulus of elasticity of a timber is only approximately related to the bending strength ratio. ASTM D245-06 addresses the modulus of elasticity by means of quality factors which are based on experimental results. The standard defines three such factors for the determination of the modulus of elasticity.

If the bending strength ratio is 55% or greater, the full clear wood modulus of elasticity can be used. If the bending strength ratio is less than 55% but equal to or greater than 45%, then a modulus of elasticity equal to 90% of the clear wood value can be used. Otherwise, the modulus of elasticity is taken as 80% of the clear wood value.

Grading Requirements of Non-U.S. Standards. The discussion above relates entirely to the indicated U.S. standards for visual grading of posts and timbers. As one would expect, other nations and regions develop their own timber production standards. Examples include BS EN 16737 *Structural Timber – Visual Strength Grading of Tropical Hardwood* (BSI 2016) from the British Standards Institution and the *Guyana Timber Grading Rules for Hardwoods* (GFC 2002) from the Guyana Forestry Commission.

The grading requirements in these and other standards do not agree exactly with those shown in Table 3.4. For example, the GFC (2002) requirements for grade GR02 hewn squares are comparable to No. 1 posts and timbers. The BSI (2016) grading requirements are likewise similar, but not identical, to those found in the U.S. standards discussed in conjunction with Table 3.4. Developing a familiarity with local standards is advised for optimum understanding, either when applying this material to work in other locales or when sourcing crane mats from outside the United States.

Considering both the grading requirements discussed here and information from mat vendors allows us to conclude that the country of origin of the timbers from which a particular crane mat is assembled does not have a significant effect on the strength and behavior of the mat. Although the precise grading characteristics will vary, the generally conservative approach taken later in this chapter in developing the suggested allowable stresses for mat design renders these differences trivial. Thus, unless specific differences in the timber qualities are known from inspection of the mats, the methods described in this handbook remain valid.

3.3.4   Application of the Timber Standards - Other Adjustments

In addition to accounting for the effects of strength-reducing defects in the lumber, D245-06 also prescribes adjustment factors that account for other differences in a piece of lumber relative to the clear green specimens. The adjustments that are applicable to crane mat design (duration of load, moisture content of the wood, and timber size) are discussed here.

Duration of Load. The allowable stresses calculated by the method defined in ASTM D245-06 are based on normal loading duration, which is defined as continuous or cumulative loading acting on the member that lasts 10 years. The application of the load duration adjustment for shorter duration loading are discussed in Section 3.5.1.

Moisture Content. The strength and stiffness of wood increases as its moisture content decreases below the fiber saturation point. However, this effect is only significant for lumber of 4 inches (102 mm) nominal thickness and less. Above that thickness, D245-06 notes that the strength increases tend to be offset by the shrinkage and seasoning defects that occur as the wood dries.

With specific consideration of crane mats, mats are continually exposed to the elements, thus preventing the timbers from attaining and remaining in a seasoned (i.e., low moisture content) state. This is consistent with the findings reported in Siegrist (2002), in which timbers from oak and mora crane mats were tested and found to have moisture contents of 68.7% and 40.7%, respectively. (Seasoned wood is that which has a moisture content no greater than 19%.) Thus, given the sizes of timbers used in crane mat construction, the moisture content factors defined in D245-06 do not apply to the determination of allowable stresses of interest here.

Size Factor. The clear specimen stress values tabulated in ASTM D2555-17a (ASTM 2017) are based on specimens that are 2" (51 mm) deep. The allowable bending stress of larger timbers determined in accordance with D245-06 must be reduced by a size factor F that is defined by Eq. 3.1.

$$F = \left(\frac{2s}{d}\right)^{1/9}$$ (3.1)

where
d =   net depth of the timber;
s =   1.0 if d is in inches; and,
         =   25.4 if d is in millimeters.

For the timber thicknesses most commonly used in crane mats, the size factor F ranges from 0.82 for 12" (300 mm) timbers to 0.86 for 8" (200 mm) timbers. The average size factor for timbers 5" (125 mm) thick, the smallest size in the posts and timbers grade, to 12" (300 mm) thick, the largest commonly available thickness, is 0.86.

3.4   TIE RODS IN TIMBER CRANE MATS

Steel tie rods are used to join the individual timbers together to form a mat. Experience shows us that the tie rods, as commonly used in the industry, are effective in holding the timbers together for handling. The issue to be discussed in this section regards the ability of the tie rods to serve as load-spreading elements. That is, if a mat made up of four timbers is supporting an outrigger and the outrigger float bears only on the two inner timbers, do the tie rods spread the applied load to the two outer timbers?

Tie rods are most commonly mild steel (ASTM A36 or another grade of similar strength) rods of diameters ranging from 3/4" to 1 3/8" (19 mm to 35 mm), with 1" (25 mm) being the most common diameter found in the crane mat dimension survey. They may be threaded only at the ends or along the entire length. As previously noted, the spacing of the tie rods was found to vary greatly, even within individual mats, although a sizable majority of the mats surveyed (about 82%) had tie rods that were spaced in the range of 15% to 30% of the overall length of the mat.

The performance of typical crane mat tie rods can be examined by way of an example. We will analyze the common arrangement of a four-timber crane mat centrally loaded by an outrigger float that bears only on the two inner timbers. The ability of the tie rods to distribute the load to the outer timbers is examined using the most common timber and tie rod dimensions. Shear and bending of the tie rods and bearing between the rods and the timbers are of significance.

EXAMPLE 3-2
    The mat, the outrigger, and the loading are shown in Fig. 3.10. Note that the ground

Outrigger load = 120,000 pounds

12" x 4' x 12' mat

22" square outrigger float

1" diameter tie rod

Ground bearing pressure = 2,500 psf

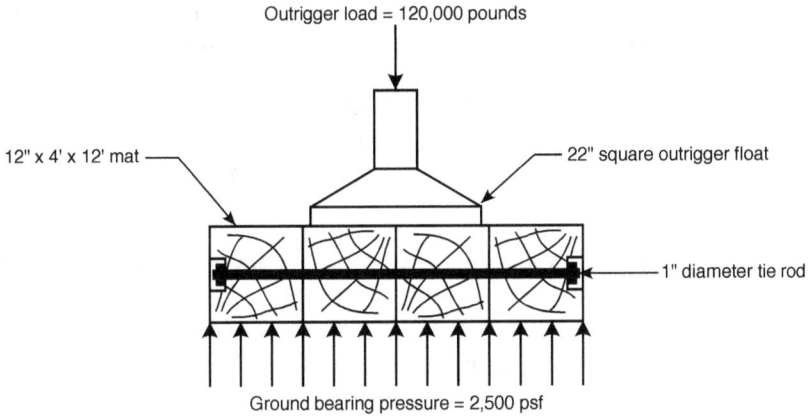

Figure 3.10 Tie Rod Loading Example

bearing pressure of 2,500 psf is only that due to the applied outrigger load. The additional ground bearing pressure due to the mat self-weight does not enter into the forces applied to the tie rods, so that bearing pressure is not included in the values shown in the figure to simplify the example.

Based on the mat dimension survey, a 12' long mat would most likely be assembled using three tie rods, one near each end and the third near the center. If a longer mat was used, say a 16' mat, then the mat would likely have four tie rods. Other pertinent dimensions are shown in the figure.

Mat bearing area = 4' x 12' = 48 ft.^{2}
Ground bearing pressure = 120,000 / 48 = 2,500 psf
Load to one outside timber = 2,500 x (1' x 12') = 30,000 pounds

The load distribution to the outside timber in such a mat would have to be carried almost entirely by the one center tie rod. Thus:

Tie rod shear = 30.0 kips
Tie rod shear area (unthreaded) = π 1.00^2 / 4 = 0.785 in.^{2}
Shear stress = 30.0 / 0.785 = 38.2 ksi
Allowable shear stress for ASTM A36 steel = 0.4 F_y = 14.4 ksi
Shear stress ratio = 38.2 / 14.4 = 2.65 > 1.00   Not Acceptable

We can see from Example 3-2 that the steel tie rods commonly used in mat construction do not have the necessary strength to distribute applied outrigger loads from the inner to the outer timbers for any but the most lightly loaded situations. In the example shown, even if two tie rods were acting to spread the load, the shear stress would have been greater than the allowable stress. Further, a bending analysis of the tie rod would show excessive bending stress in the tie rods and excessive bearing stress between the tie rods and the timbers.

Many mobile cranes of 75 tons (68 tonnes) capacity or greater have outrigger floats that are about 30 inches (760 mm) or greater in size. A float of this size will bear directly over all four timbers in a standard 12" x 4' (300 mm x 1,220 mm) mat. In this case, the tie rods only have to hold the timbers in relative position and do so by acting in tension (similar in behavior to the bars in a reinforced concrete beam). Experience shows that the tie rods are effective in this function.

3.5   STRENGTH AND STIFFNESS DESIGN VALUES FOR TIMBER MATS

This chapter has presented the results of a survey of the dimensions and conditions of timber crane mats in service (Section 3.1), a survey of crane mat vendors to determine the species of wood used in the manufacture of mats (Section 3.2), a discussion of industry standards applicable to the determination of timber strength (Section 3.3), and a brief analysis of the contributions of the tie rods used in mat construction (Section 3.4). In this section, we will use that information to develop allowable stresses and a modulus of elasticity that are appropriate for use in the analysis of timber crane mats.

3.5.1   Allowable Stresses Based on NDS (AWC 2016) – Common Species

The *National Design Specification*® *for Wood Construction* (AWC 2016), commonly referred to simply as the *National Design Specification* or the NDS, along with its previous editions, has long been used as a convenient source of allowable stresses for crane mat design. In this section, allowable stresses for the common (in the U.S.) species used for crane mats are developed based on the values and provisions of AWC (2016). (The term "common species" is used to collectively identify the various hardwood and softwood species first listed in Table 3.3, except for azobe and mora. These woods can treated as a group since their strength properties are relatively similar.)

Basic allowable stresses for a variety of timber species commonly used for crane mat construction are assembled in Table 3.5. These allowable stresses are for No. 1 posts and timbers, the grade that has typically been assumed applicable to the timbers used in crane mat construction and which the timber dimension and condition survey has shown is appropriate.

Table 3.5 includes a species group labeled "mixed hardwoods." This is a group of hardwoods that includes beech, birch, and a variety of hickory species. The indicated properties were listed in earlier versions of the NDS up to and including the 1986 edition (NFPA 1986). Although no longer listed in the NDS, the group and its properties are shown here for reference. It is also noted that the method of determining the shear strength of timbers was revised beginning with the 2001

TABLE 3.5 Typical Allowable Stresses from the NDS (AWC 2016)

(a) Values in U.S. Customary Units:

Species	F_b, psi	F_v, psi	$F_{c\perp}$, psi	E, psi	Density, pcf
Hickory [1]	1,250	180	715	1,500,000	50
Mixed Hardwoods [1,4]	1,250	180	715	1,500,000	50
Mixed Oak [1]	1,000	155	800	1,000,000	47
Northern Red Oak [1]	1,200	205	885	1,300,000	47
White Oak [1]	1,050	205	800	1,000,000	50
Douglas Fir-Larch [2,3]	1,200	170	625	1,600,000	32
Western Hemlock [1]	1,050	170	410	1,400,000	32

(b) Values in SI Units:

Species	F_b, kPa	F_v, kPa	$F_{c\perp}$, kPa	E, kPa	Density, kg/m^3
Hickory [1]	8,600	1,240	4,950	10,300,000	801
Mixed Hardwoods [1,4]	8.600	1,240	4,950	10,300,000	801
Mixed Oak [1]	6,900	1,070	5,500	6,900,000	753
Northern Red Oak [1]	8,250	1,410	6,100	9,000,000	753
White Oak [1]	7,250	1,410	5,500	6,900,000	801
Douglas Fir-Larch [2,3]	8,250	1,170	4,300	11,000,000	513
Western Hemlock [1]	7,250	1,170	2,850	9,700,000	513

Values are from the *National Design Specification® for Wood Construction* (AWC 2016), Table 4D, for Posts & Timbers grade No. 1, except as otherwise noted
Density is estimated for as-used moisture content
[1] NeLMA grading rules
[2] WCLIB grading rules
[3] NLGA grading rules
[4] Refer to text for details of this species group

edition of the NDS. The value of F_v listed in NFPA (1986) was 90 psi (620 kPa), which is comparable to 180 psi (1,240 kPa) using the current methodology. The larger value is shown in Table 3.5 to permit easy comparison to the allowable shear stresses for the other species.

The actual species of the timbers normally will not be known to the mat user, so we must consider all of the values shown in Table 3.5 and select basic allowable stresses that can be applied universally. We can also consider increasing those allowable stress values based on the time duration of loading following the procedure defined in the NDS (AWC 2016).

The allowable stresses given in the NDS are based on a normal load duration, which is defined in ASTM D245-06 as continuous or cumulative loading lasting 10 years, a condition that is appropriate for use in the design of conventional buildings or other permanent structures. Normal load duration by this definition is clearly not the loading condition applicable to the design of mobile crane supports, so investigation of the NDS load duration factor is in order. For this, we must make some educated assumptions regarding the use of a crane mat with respect to time and loading.

Crane mats are exposed to the elements and are subject to abuse as a result of their normal use. Consequently, a mat will degrade relatively quickly. Discussions with lifting contractors around the United States indicate that the upper bound useful life of a timber crane mat is about four years. This tends to be the case in northern regions. In the southern U.S., where damper conditions and a greater problem with termites exist, the useful life of a mat may be only one or two years. For this analysis, we will conservatively use the upper bound value of four years (note that the assumption of a longer life yields more conservative results for the load duration analysis being performed here).

A crane mat isn't used continuously. A significant portion of its life is spent in storage. A reasonable upper bound assumption for crane mat usage is that about 50% of the mat's life is spent under some level of loading and the remaining 50% is spent in storage. Of that 50% under load, not every use will see high loads that bring the stresses in the mat up near the allowable values. If we assume that a contractor will try to match mat sizes with the crane service, we may assume that about half of the days on a job site will result in loading that will result in high stresses. And last, we know that the peak loads from an outrigger or under a crawler track occur only briefly, as the crane swings or booms out with a load. We will assume that the duration of high loading averages about half an hour for each day that high loading occurs. The remaining 23 1/2 hours of each day in use will see some level of loading (the crane sits on the mats around the clock, even when not making lifts, for all of the time in which the mat is in service), which we will identify here as "medium" loading. This medium loading will be assumed to be equal to about 50% of the peak loading. It is this set of loading conditions that we will use in application of the load duration adjustment factor.

It is clear from the list of assumptions outlined in the previous paragraphs that we do not have hard statistical data on crane mat use, loading, and length of useful life. In the absence of data, we must proceed by making experience-based assumptions. The figures in the preceding paragraphs are based on the author's experience and on estimates provided over the years by many other lift specialist engineers and contractors. What now follows is a calculation of allowable stresses for crane mat design based on these values.

Here we note the most important aspect of applying the load duration factor to the calculation of allowable stresses: The time duration of loading is *cumulative*. We cannot look at a one-hour occurrence of a peak load and say that the duration is one hour. Rather, we must look at the lifetime use of the mat and tally up all of those peak load events to arrive at the true duration of the particular magnitude of loading under consideration.

The NDS prescribes a load duration factor C_D by which allowable stresses for normal loading duration can be increased for loadings of shorter durations. C_D is defined by values in a table and on a graph. This information can be distilled into equation form as Eq. 3.2.

$$C_D = \frac{4}{3} D^{-1/30} \qquad (3.2)$$

where D is the cumulative loading duration in days.

Based on the discussion above, we can analyze the loading conditions and durations that a typical crane mat will experience over its life as follows. (As noted above, the loading and load durations upon which this analysis is based are based on the author's experience and anecdotal information collected from other lift specialist engineers and crane users. The analysis following can serve as a model in the future should a more rigorous study of crane loading be conducted and published for use by the industry.)

CRANE MAT LOAD DURATION ANALYSIS
Total life = 4 years at 365.25 days per year = 1,461.000 days

Case 1 - Use Under Peak Loads
Number of days in any use = 50% of 1,461.000 = 730.500 days
Number of days with peak loads = 50% of 730.500 = 365.250 days
Time per day under peak loading = 0.5 hour
Hours under peak loading = 0.5 hour per day x 365.250 days = 182.625 hours
Equivalent number of days under peak loading = 182.625 / 24 hours per day = 7.609 days
Apply Eq. 3.2 to compute C_D for peak loading condition:

$$C_D = \frac{4}{3} 7.609^{-1/30} = 1.246$$

Case 2 - Use Under Medium Loads
Number of days in any use = 730.500 days (from Case 1)
Number of days with medium loads = 730.500 - 7.609 = 722.891 days
Apply Eq. 3.2 to compute C_D for medium loading condition:

$$C_D = \frac{4}{3} 722.891^{-1/30} = 1.071$$

Combine the two values of C_D in accordance with NDS Appendix B:
Define the peak load condition as 100.
Define the medium load condition as 50% of the peak load condition = 50
Peak Load Factor = 100 / 1.246 = 80.25
Medium Load Factor = 50 / 1.071 = 46.70
Peak Load Factor > Medium Load Factor ⇒ Peak Load value of C_D controls:
   C_D = 1.246

The last factor to be considered is the ratio of the actual timber section properties to the properties based on nominal dimensions. Crane mats are commonly evaluated using the nominal dimensions (see Chapters 4 and 5 for more on this subject). Since the average actual dimensions are slightly less than the nominal dimensions (but not as much less as occurs in lumber used for construction that has been surfaced on all four sides), the allowable stresses should be adjusted by this ratio to assure that the final calculated strength is representative of the actual strength provided by the mat. The timber dimension survey results in Table 3.2 show that the ratio of actual section modulus to nominal section modulus is 0.922 and the ratio of actual cross sectional area to nominal area is 0.961 (based on the findings for the 12" x 12" timbers).

Looking at the allowable stress values in Table 3.5 and considering the species of timbers most commonly used in the manufacture of crane mats, we can see that reasonable allowable stress values for normal duration loading are 1,200 psi (8,250 kPa) for bending and 170 psi (1,170 kPa) for shear. Thus, application of the provisions of the NDS (AWC 2016) using the value of $C_D = 1.246$ and the section property ratios from Table 3.2 gives us the following allowable stresses for bending and shear.

$$F_b = 1,200 \text{ x } 1.246 \text{ x } 0.922 = 1,400 \text{ psi}$$

$$F_v = 170 \text{ x } 1.246 \text{ x } 0.961 = 200 \text{ psi}$$

The calculated values of F_b and F_v shown above are rounded in accordance with D245-06 to arrive at these final values for design use.

As noted in the NDS, the load duration adjustment factor is not applied to compression perpendicular to the grain. Further, the applied compression stress is a function of the footprint of the crane (crawler track or outrigger float dimensions), so the results of the dimension survey do not affect this allowable stress.

The last value to address is the modulus of elasticity, E. Given the values in Table 3.5, a value of 1,300,000 psi (9,000,000 kPa) appears to be appropriate for the common timber hardwood and softwood species that are addressed here. However, there is one important point that must be considered. As noted above for the development of the bending and shear allowable stresses, mat calculations are typically performed using the nominal dimensions of the timbers. Therefore, we should adjust the NDS value of E from Table 3.5 by the ratio of the actual moment of inertia (the average value based on the dimension survey findings) to the nominal moment of inertia.

The ratio of actual moment of inertia to nominal moment of inertia is shown in Table 3.2 as being equal to 0.923 for the most common size of crane mat timbers [12" x 12" (300 mm x 300 mm)]. This gives us an adjusted value of E equal to 1,300,000 x 0.923 = 1,199,900 psi. This value can be rounded off to 1,200,000 psi (8,300,000 kPa) for practical use in mat deflection calculations. And as noted in the NDS, the load duration adjustment factor is not to be applied to the modulus of elasticity.

3.5.2 Allowable Stresses Based on the NDS (AWC 2016) Method – Tropical Hardwoods

The method based on AWC (2016) developed in the previous section can be used to arrive at suitable allowable stresses for the design of crane mats built with timbers of other species, such as tropical hardwoods. Since these species are not addressed in the NDS, we don't have normal load duration allowable stresses to use as a starting point. Therefore, the following derivation is proposed based on the use of clear wood strength values.

Clear wood strength values are shown in Table 3.6 for the timber species listed in Table 3.5. These values are based on the tables in ASTM D2555-17a (ASTM 2017) for those species listed in the standard and estimated from the values in D2555-17a for the mixed hardwoods and mixed oak groups. There are seven tropical hardwoods used in crane mat construction, particularly from European suppliers. These are azobe (also called ekki; native to West Africa), dahoma (also called dabema; native to West Africa), eucalyptus (native to Australia, cultivated in South America), greenheart (native to northeastern South America), mora (native

TABLE 3.6 Clear Wood Stresses for the Common Species

(a) Values in U.S. Customary Units:

Species	Modulus of Rupture, psi	Shear Strength, psi	Comp. Perp. to Grain, psi	E_c, psi
Hickory	9,060	1,032	1,250	1,289,000
Mixed Hardwoods	9,050	1,030	1,250	1,289,000
Mixed Oak	7,000	900	1,110	1,245,000
Northern Red Oak	8,300	1,214	987	1,353,000
White Oak	8,300	1,249	1,109	1,246,000
Douglas Fir-Larch	7,540	922	773	1,613,000
Western Hemlock	6,270	933	632	1,038,000

(b) Values in SI Units:

Species	Modulus of Rupture, kPa	Shear Strength, kPa	Comp. Perp. to Grain, kPa	E_c, kPa
Hickory	62,500	7,120	8,620	8,887,000
Mixed Hardwoods	62,400	7,100	8,620	8,887,000
Mixed Oak	48,300	6,210	7,650	8,584,000
Northern Red Oak	57,200	8,370	6,810	9,329,000
White Oak	57,200	8,610	7,650	8,591,000
Douglas Fir-Larch	52,000	6,360	5,330	11,121,000
Western Hemlock	43,200	6,430	4,360	7,157,000

to northeastern South America), tonka (also called cumaru; native to northeastern South America), and wamara (also called Guyana rosewood; native to regions from southern Mexico through northeastern South America). Corresponding clear wood strength values for these three species are listed in Table 3.7.

The properties shown in Tables 3.6 and 3.7 are for green wood, not seasoned timbers. As discussed on page 103, this is closer to the condition we must expect for crane mat timbers. Therefore, these are the appropriate properties for use in this strength evaluation.

By comparing the clear wood properties in Table 3.6 to the allowable stresses for crane mat design developed above for the common timber species, we see that the average ratio of the allowable bending stress to the modulus of rupture and the average ratio of the allowable shear stress to the shear strength are both on the order of 0.2. These ratios can be applied to the clear wood strength values of other species, such as those listed in Table 3.7, to arrive at allowable stresses that

TABLE 3.7 Clear Wood Stresses for Tropical Hardwoods

(a) Values in U.S. Customary Units:

Species	Modulus of Rupture, psi	Shear Strength, psi	Side Hardness, lbs.	E_c, psi	Density, pcf
Azobe [1]	16,900	2,040	2,890	2,160,000	60
Dahoma [2]	11,000	1,100	1,320	1,430,000	40
Eucalyptus [1]	9,900	1,320	1,290	1,480,000	40
Greenheart [1]	19,300	1,930	1,880	2,470,000	60
Mora [1]	12,600	1,400	1,450	2,330,000	55
Tonka [2]	19,290	1,930	2,200	2,690,000	60
Wamara [2]	21,400	2,140	3,325	2,480,000	60

(b) Values in SI Units:

Species	Modulus of Rupture, kPa	Shear Strength, kPa	Side Hardness, N	E_c, kPa	Density, kg/m³
Azobe [1]	116,500	14,100	12,900	14,900,000	960
Dahoma [2]	75,800	7,600	5,900	9,900,000	640
Eucalyptus [1]	68,300	9,100	5,700	10,200,000	640
Greenheart [1]	133,100	13,300	8,400	17,000,000	960
Mora [1]	86,900	9,700	6,400	16,100,000	880
Tonka [2]	133,000	13,300	9,800	18,500,000	960
Wamara [2]	147,500	14,800	14,800	17,100,000	960

[1] Properties from FPL (2010)
[2] Properties other than shear strength from FPL (1984); shear strength estimated from similar species (see text for details)

are suitable for crane mat design. Thus, expressions for the allowable bending and shear stresses suitable for mat design can be written as Eqs. 3.3 and 3.4, respectively.

$$F_b = 0.22 \, S_b MR_c \tag{3.3}$$

$$F_v = 0.21 \, S_v V_c \tag{3.4}$$

where

S_b	=	ratio of actual S_x to nominal S_x;
	=	0.92 when using nominal dimensions of rough-sawn timbers;
	=	1.00 when using actual timber dimensions;
MR_c	=	clear wood modulus of rupture;
S_v	=	ratio of actual cross-sectional area to nominal cross-sectional area;
	=	0.96 when using nominal dimensions of rough-sawn timbers;
	=	1.00 when using actual timber dimensions;
V_c	=	clear wood shear strength.

The values of the dimensional factors S_b and S_v are based on the results of the timber dimension survey reported in Section 3.1. Note that 0.22 S_b and 0.21 S_v both equal 0.20, which agrees with the relationship discussed on page 112.

The values for mora can be used as an example. Given the clear wood strength values shown in Table 3.7, we obtain design values of 0.22 x 0.92 = 0.20 and 0.20 x 12,600 = 2,520 psi (17,380 kPa) for the allowable bending stress and 0.21 x 0.96 = 0.20 and 0.20 x 1,400 = 280 psi (1,940 kPa) for the allowable shear stress. The bending stress value should be rounded down to 2,500 psi (17,300 kPa) for practical use in crane mat design and analysis. The shear stress value of 280 psi may be used as computed.

For reference, these allowable stresses may be somewhat conservative. The test results reported in Siegrist (2002) showed mora as having a bending strength just over twice that of oak. Unpublished results of tests performed at the University of Alberta, also in 2002, on Douglas fir and mora timbers showed the shear strength of mora to be almost 50% greater than that of Douglas fir. By comparison, the allowable stresses suggested above for mora are 79% and 40% greater than those developed for the common hardwood and softwood species in bending and shear, respectively. Given the very limited scope of the tests done at Lehigh and Alberta, it is prudent to use the lower values developed here.

One value is lacking in FPL (2010) and FPL (1984) for some of the tropical hardwoods. A compressive stress perpendicular to the grain is not given. Rather, the compressive strength perpendicular to the grain is expressed as a side hardness. By comparing the side hardness values in FPL (2010) for the hardwood species in Table 3.5 to the specified allowable compressive stress, we find that the allowable stress in psi is approximately equal to 0.75 times the side hardness in pounds (Eq. 3.5a) or the allowable stress in kPa is approximately equal to 1.16 times the side hardness in newtons (Eq. 3.5b).

$$F_{c\perp} = 0.75H_c \qquad\qquad (3.5a)$$

$$F_{c\perp} = 1.16H_c \qquad\qquad (3.5b)$$

where
$F_{c\perp}$ =   allowable compressive stress perpendicular to the grain, psi (Eq. 3.5a) or kPa (Eq. 3.5b);
H_c =   clear wood side hardness, pounds (Eq. 3.5a) or newtons (Eq. 3.5b)

The modulus of elasticity for timbers of other species can be calculated from the clear wood value E_c shown in Table 3.7, the quality factor of 0.90 for No. 1 posts and timbers from D245-06 Table 5, the adjustment factor of 0.94 from D245-06 Table 8, and the moment of inertia ratio of 0.923 discussed in the preceding section. This solution is shown in Eq. 3.6.

$$E = \frac{0.90 \times 0.923}{0.94} E_c = 0.88 E_c \qquad\qquad (3.6)$$

If the section properties of the mat are calculated using the actual timber dimensions, then Eq. 3.6 should be modified by removing the 0.923 term.

The calculation of allowable stresses from clear wood stresses requires rounding of the final values in accordance with ASTM D245-06 (ASTM 2011). Conversions of stress values that are developed in U.S. customary units (USCU) to values in SI units likewise requires rounding, here in accordance with ASTM SI 10-2016 (ASTM 2016). A summary of the rounding requirements from these two standards are given in Appendix 3.

Allowable stresses suggested for use in crane mat design are assembled in Table 3.8. The derivation of these values, as discussed in this section, are based in part on the assumption that the relationship of the actual timber dimensions to the nominal dimensions are as determined in the timber dimension survey. Some mats made with tropical hardwoods use timbers that are finished on four sides (that is, for example, an 8" x 8" timber is actually 7 5/8" x 7 5/8"). In such a case, the allowable bending stresses in Table 3.8 may be increased by 8.5%, the allowable shear stresses by 4.1% and the modulus of elasticity by 8.3%, but the mat section properties then must be calculated using the actual finished dimensions of the timbers, not the nominal dimensions.

In summary, the derivations presented in Sections 3.5.1 and 3.5.2 give us reasonable and practical values for allowable stresses and the modulus of elasticity for timber crane mat design based on established timber design practice and mechanical properties from recognized sources for both common hardwood and softwood mats and for mats made of other species, such as tropical hardwoods. The procedure used in Section 3.5.2 can be used to develop allowable stresses

TABLE 3.8 Suggested Allowable Stresses for Timber Crane Mats Derived using the NDS (AWC 2016) Methodology

(a) Values in U.S. Customary Units:

Species	F_b, psi	F_v, psi	$F_{c\perp}$, psi	E, psi	Density, pcf
Common Species	1,400	200	750	1,200,000	50
Azobe (Ekki)	3,400	410	2,170	1,900,000	60
Dahoma (Dabema)	2,200	220	990	1,300,000	40
Eucalyptus	2,000	265	970	1,300,000	40
Greenheart	3,850	385	1,410	2,200,000	60
Mora	2,500	280	1,090	2,100,000	55
Tonka (Cumaru)	3,850	385	1,650	2,400,000	60
Wamara	4,300	430	2,495	2,200,000	60

(b) Values in SI Units:

Species	F_b, kPa	F_v, kPa	$F_{c\perp}$, kPa	E, kPa	Density, kg/m^3
Common Species	9,650	1,380	5,150	8,300,000	800
Azobe (Ekki)	23,500	2,820	15,000	13,100,000	960
Dahoma (Debema)	15,300	1,530	6,800	9,000,000	640
Eucalyptus	13,600	1,840	6,600	9,000,000	640
Greenheart	26,500	2,650	9,800	15,200,000	960
Mora	17,300	1,940	7,400	14,500,000	880
Tonka (Cumaru)	26,500	2,650	11,400	16,600,000	960
Wamara	29,600	2,960	17,200	15,200,000	960

for other species using clear wood values from FPL (2010) or other recognized sources when the need arises.

For reference, the author has been using the allowable stresses shown in Table 3.8 for the common wood species for crane mat design for many years with success, both in the technical sense and with respect to acceptance by many other lift specialist engineers and contractors in the construction industry throughout the United States.

3.5.3   Allowable Stresses Based on the Timber Survey

The allowable stresses developed in Sections 3.5.1 and 3.5.2 are based, as stated, on the provisions of the *National Design Specification* (AWC 2016). The process used is based on one fundamental assumption, which is that the timbers that make up a crane mat generally conform to the conditions consistent with the No. 1 grade for posts and timbers. Given the information that we have in hand from the crane mat dimension survey, as outlined in Section 3.1, we can approach the development of allowable stresses from a second direction. In this section, we will demonstrate the calculation of the values of allowable stresses by using the survey data and the strength ratios defined in ASTM D245-06. In the interest of brevity, we will run through the procedure only for the allowable bending stress of hickory, the first species listed in Table 3.5. The same procedure can be applied to the other stresses and other species.

The derivation of the NDS allowable bending stress is shown in Table 3.9 using hickory as an example. Working down through the middle column in the table, we start with the average clear wood bending strength as represented by the modulus of rupture and the standard deviation of that property, both taken from ASTM D2555-17a. The third line calculates the 5% exclusion limit S of the bending strength, given by Eq. 3.7 (see Note 10 of D2555-17a). (The 5% exclusion limit is the level of strength below which 5% of the strength values are expected to fall and corresponds to a selected probability point from the frequency distribution of strength values.)

$$S = \mu - 1.645\sigma \qquad (3.7)$$

where

μ =   the average (mean) value of the strength property; and,
σ =   the standard deviation of that property.

The next lines show the strength ratios that affect allowable bending strength. For posts and timbers, these are the strength ratios for knots and the slope of grain. The values shown are taken from ASTM D245-06 for the limits of these characteristics permitted by the relevant grading standards (see Table 3.4). These

TABLE 3.9 Derivation of Allowable Bending Stresses

Quantity	No. 1 Posts and Timbers	Hickory using Survey Data
Clear Wood Bending Strength, psi	9,060	9,060
Standard Deviation, psi	1,450	1,450
5% Exclusion Bending Strength, psi	6,675	6,675
Knot Strength Ratio	0.50	0.66
Slope of Grain Strength Ratio	0.61	0.69
Controlling Strength Ratio	0.50	0.66
Size Factor	0.86	0.86
Adjustment Factor	2.30	2.30
Calculated Allowable Bending Stress, psi	1,248	1,647
NDS Allowable Bending Stress, psi	1,250	

strength ratios are not additive. Rather, the lowest value controls, shown in the table as 0.50 (the knot strength ratio). The size factor value of 0.86 is based upon a range of timber sizes as discussed on page 104. Last, the adjustment factor is a structural design factor defined in Table 8 of D245-06, which is equal to 2.30 for hardwoods and 2.10 for softwoods. The calculated allowable bending stress is the 5% exclusion value multiplied by the controlling strength ratio and the size factor, and divided by the adjustment factor. The result, 1,248 psi, is rounded off to 1,250 psi, which is the allowable bending stress given by NDS for hickory for normal loading.

This process can be repeated for other species groups. The key, of course, is to start with the strength properties that are appropriate for the species or species group of interest. ASTM D2555-17a discusses how strength properties of individual species are combined to develop properties of groups. However, given the convenience and construction industry acceptance of using NDS-based allowable stresses for crane mat design, repeating this exercise will rarely be necessary in practice.

The right column in Table 3.9 repeats this process, except that the strength ratios for knots and slope of grain are based on the observations made in the crane mat dimension and condition survey. Here, both the knot sizes (e.g., Fig. 3.6) and the slope of grain (e.g., Fig. 3.8) fell within the limits for Select Structural. Thus, the strength ratios shown in the right column of Table 3.9 are based on that higher grade. The resulting allowable bending stress is found to be 1,647 psi, 32% greater than the NDS allowable bending stress value of 1,250 psi. This 32% difference is discussed further in the next section.

The strength ratios used in the determination of allowable shear stresses are those for checks, shake, and splits. As discussed in Section 3.3.3, the strength ratios for these defects are the same for all lumber grades. Therefore, the findings of the timber dimensions and condition survey do not affect the determination of shear strength.

It is obvious from some of the photographs in this chapter that some crane mat timbers have splits that exceed the lengths permitted by the grading standards (Table 3.4). The bending and shear strength of such a timber can be calculated giving consideration to the other defects present (or not present) in the vicinity of these splits. Doing so shows that the shear and bending capacities based on the allowable stresses in Table 3.8 can still be achieved in the presence of such overlength splits.

As noted in the beginning of this chapter, the dimension and condition survey reported in Section 3.1 only address crane mats manufactured and used throughout the United States. Sawmill practices, available species, and mat manufacturing methods employed elsewhere around the world may vary, possibly significantly, from those upon which this work is based. Thus, the discussions and recommendations made may not be directly suitable for use outside of the U.S. However, the process shown here for the development of design values can be applied using data collected in other regions.

3.5.4   Limit State Stresses for the Common Species

Under normal circumstances, crane mat design will proceed using the allowable stresses shown in Table 3.8. In unique situations, however, the need may arise to check a design or to research a particular concept using limit state stresses of the timbers. Practical values of the bending and shear limit state stresses can be derived easily using the values developed in the preceding sections.

Consider first the species listed in Table 3.3 and used throughout this chapter. The allowable stresses for crane mat design were derived from the values in AWC (2016) by adjusting for the time duration of use typical to crane mat service and for the ratio of actual section properties to the nominal section properties that are commonly used in calculations. The limit state stresses of these species can be calculated in a similar fashion. Here, the AWC (2016) allowable stresses are multiplied by the adjustment factor (structural design factor) K, by the load duration factor C_D, and by the ratio of actual to nominal section properties.

The value of C_D used in the derivation of allowable stresses for design is based on an assumed usage over the life of a crane mat. Further, the basis for the load duration factor lies in the behavior of wood where the ultimate strength of a member is a function of the time under load. Since an overload can occur any time within the life of the mat, the same value of C_D used in developing the allowable stresses should be used in the development of the limit state stresses.

Last, we must also consider the findings of the timber survey and their effect on bending strength, as discussed above in conjunction with Table 3.9. As previously noted, the allowable stress calculation procedure defined in D245-06 and the values tabulated in the NDS do not account for any degradation due to decay, rot, or the like. Thus, it is not reasonable to assume that the bending strength of a crane mat in practice is actually 32% greater than the strength values upon which the allowable stresses given in Table 3.8 are based. Some of this extra strength will be lost to degradation during the useful life of the mat. Should a limit state strength analysis of timber crane mats be necessary, one may reasonably assume that a fraction of this additional strength is available at any random time during the life of a mat. The use of bending limit state values 15% greater than those calculated for the No. 1 grade is suggested. Thus, the limit state stresses are calculated using Eq. 3.8 for bending and Eq. 3.9 for shear.

$$F_{b \text{ limit state}} = F_{b \text{ NDS}} \; K \; C_D \frac{S_{x \text{ actual}}}{S_{x \text{ nominal}}} 1.15 \qquad (3.8)$$

$$F_{v \text{ limit state}} = F_{v \text{ NDS}} \; K \; C_D \frac{A_{\text{actual}}}{A_{\text{nominal}}} \qquad (3.9)$$

All of the necessary values are assembled in Table 3.10 for the limit state stresses in bending and in Table 3.11 for the limit state stresses in shear. The basic allowable stresses are from the NDS (AWC 2016), the adjustment factors

TABLE 3.10 Timber Limit State Stresses in Bending Based on Timber Condition

Species	F_b, psi	Adjust. Factor K	C_D	Actual S_x / Nominal S_x	Survey Factor	Limit State F_b, psi
Hickory	1,250	2.3	1.246	0.922	1.15	3,800
Mixed Hardwoods	1,250	2.3	1.246	0.922	1.15	3,800
Mixed Oak	1,000	2.3	1.246	0.922	1.15	3,050
Northern Red Oak	1,200	2.3	1.246	0.922	1.15	3,650
White Oak	1,050	2.3	1.246	0.922	1.15	3,200
Douglas Fir-Larch	1,200	2.1	1.246	0.922	1.15	3,350
Western Hemlock	1,050	2.1	1.246	0.922	1.15	2,900
					Average Value, psi	3,400
					Average Value, kPa	23,400

TABLE 3.11 Timber Limit State Stresses in Shear Based on Timber Condition

Species	F_v, psi	Adjust. Factor K	C_D	Actual A / Nominal A	Survey Factor	Limit State F_v, psi
Hickory	180	2.3	1.246	0.961	1.00	495
Mixed Hardwoods	180	2.3	1.246	0.961	1.00	495
Mixed Oak	155	2.3	1.246	0.961	1.00	425
Northern Red Oak	205	2.3	1.246	0.961	1.00	565
White Oak	205	2.3	1.246	0.961	1.00	565
Douglas Fir-Larch	170	2.1	1.246	0.961	1.00	425
Western Hemlock	170	2.1	1.246	0.961	1.00	425
					Average Value, psi	485
					Average Value, kPa	3,340

are $K = 2.3$ for hardwoods and $K = 2.1$ for softwoods per ASTM D245-06 (ASTM 2011), the value of C_D is taken as 1.246 as previously calculated, the ratios of actual to nominal section properties are as discussed in conjunction with Table 3.2, and the "survey factor" of 1.15 applies the 15% increase in bending strength discussed in the preceding paragraph. As this increase applies to bending strength only, the survey factor does not appear in Eq. 3.9. The ratios of actual to nominal section properties are based on the results for the mats with 12" x 12" timbers only. This was done due to the vastly greater number of specimens of this size that were reported in the survey, along with the large geographical diversity of the reporting survey participants.

Eqs. 3.8 and 3.9 are used to populate the rightmost columns in Tables 3.10 and 3.11, respectively. The limit state stresses calculated for each species or species group are then averaged to arrive at average limit state stresses of 3,400 psi (23,400 kPa) in bending and 485 psi (3,340 kPa) in shear for the common species of hardwoods and softwoods used in crane mat construction in the U.S. For reference, these average limit state values are 2.43 times the allowable stress values in both bending and shear shown in Table 3.8, thus providing an indication of the actual structural design factor provided by the recommended allowable stresses.

A final note must be made about the values developed in Tables 3.10 and 3.11. The calculation of these limit state stresses include factors that account for the differences in actual dimensions with respect to the nominal dimensions of timbers. These factors are based on the dimension survey conducted by the author (Section 3.1) and on the load duration factor that is specific to crane mat use (Section 3.5.1). Thus, these factors are unique to the timbers used in the construction of crane mats. Limit state or ultimate strength values for timbers that are reported in references that are applicable to the design of other types of structures (e.g. building structures) may not agree with the values given here, simply because the assumed conditions of use differ.

3.5.5   Limit State Stresses for Tropical Hardwoods

The calculations performed in Section 3.5.4 for the common species can be used to develop limit state stresses for the tropical hardwoods. A comparison of the limit state stresses to the clear wood stresses show that the limit state bending stress is, on average, 42.0% of the modulus of rupture and the limit state shear stress is, again on average, 47.0% of the clear wood shear strength. These percentage values can be used to derive reasonable limit state stresses in bending and shear for other species, such as the tropical hardwoods, for which the NDS does not specify allowable stresses.

The limit state stresses in bending and in shear are simply the clear wood modulus of rupture MR_c or shear strength V_c multiplied by 0.420 or 0.470, respectively. The clear wood values and the limit state stresses thus calculated and appropriately rounded for the seven tropical hardwoods are assembled in Table 3.12 for bending and Table 3.13 for shear.

3.5.6   Lower Strength Timber Species

As noted at the end of Section 3.2, some crane mat vendors use timbers of species that have strength characteristics that are lower than those of the common species first listed in Table 3.3 upon which the discussion in this section are based. The species of which the author is aware are listed in Table 3.14, along with the primary strength values from ASTM D2555-17a (ASTM 2017). By comparison to the strength values shown in Table 3.6 for the common species, we can see that these species are significantly lower in bending, shear, and compression strength.

TABLE 3.12 Tropical Species Limit State Stresses in Bending Based on Timber Condition

Species	Limit State F_b / MR_c	Modulus of Rupture		Limit State F_b	
		psi	kPa	psi	kPa
Azobe (Ekki)	0.420	16,900	116,500	7,100	49,000
Dahoma (Debema)	0.420	11,000	75,800	4,620	31,900
Eucalyptus	0.420	9,900	68,300	4,160	28,700
Greenheart	0.420	19,300	133,100	8,110	55,900
Mora	0.420	12,600	86,900	5,290	36,500
Tonka (Cumaru)	0.420	19,290	133,000	8,100	55,800
Wamara	0.420	21,400	147,500	8,990	62,000

TABLE 3.13 Tropical Species Limit State Stresses in Shear Based on Timber Condition

Species	Limit State F_v / V_c	Shear Strength		Limit State F_v	
		psi	kPa	psi	kPa
Azobe (Ekki)	0.470	2,040	14,100	960	6,620
Dahoma (Debema)	0.470	1,100	7,600	515	3,550
Eucalyptus	0.470	1,320	9,100	620	4,270
Greenheart	0.470	1,930	13,300	905	6,240
Mora	0.470	1,400	9,700	660	4,550
Tonka (Cumaru)	0.470	1,930	13,300	905	6,240
Wamara	0.470	2,140	14,800	1,005	6,930

(Table 3.14 shows only USCU values since this table is for comparison only, not for design use.)

Given the generally lower strength values, particularly for cottonwood and yellow poplar, the allowable stresses developed in this chapter for crane mat design are not compatible with mats made of these species.

TABLE 3.14 Clear Wood Stresses of Lower Strength Species

Species	Modulus of Rupture, psi	Shear Strength, psi	Comp. Perp. to Grain, psi	E_c, psi
Cottonwood	5,260	682	354	1,013,000
Hackberry	6,480	1,070	676	954,000
Sweet Gum	7,110	992	626	1,201,000
Sycamore	6,470	996	622	1,065,000
Yellow Poplar	5,950	792	470	1,222,000

The best course of action is to avoid mats made of these species, thus removing all concern about the applicability of the allowable stresses and design methods. If the only mats available are understood to be made of timbers of these (or other low strength) species, then the allowable stresses given in Table 3.8 for the common species should be reduced by an appropriate percentage for use with these (or any other) species of lower strength. The actual percentage reduction must be determined by the lift planner/engineer based on the strength properties of the actual species used.

It is noted that the species listed in Table 3.14 are not covered by Table 4D of the *National Design Specification* (AWS 2012), so development of allowable stresses for crane mat design using NDS values, as was done for the common species, is not an option. The method developed in Section 3.5.2 for the determination of allowable stresses for the tropical hardwoods may be useful when evaluating mats made of timbers of the species listed in Table 3.14.

3.6   FABRICATED STEEL MAT PROPERTIES

Many large cranes develop support reactions, either outrigger loads or crawler track bearing pressures, that are simply too great for the load spreading ability of timber crane mats to disperse effectively. One answer to this problem is fabricated steel crane mats.

Steel mats for crawler cranes can be very simple. A design commonly used employs a series of I-shape rolled beams, either W or S shapes, or structural tubes arranged parallel to one another, just like the timbers in a conventional crane mat. The beams or tubes either can be welded directly to one another along the edges of their flanges or with flare-bevel welds in the corners (Fig. 3.11*a*) or they can be spaced apart with cover plates welded top and bottom to form a built-up section (Fig. 3.11*b*). In use, the mats normally will be positioned such that the beams are perpendicular to the length of the crawler track. Thus, the mat can be designed for shear and bending in one direction only, same as the design of a timber crane mat. If the beams are spaced apart, the bottom cover plate must also be designed for the bending stresses due to the ground bearing pressure on the plate as it spans between adjacent beams.

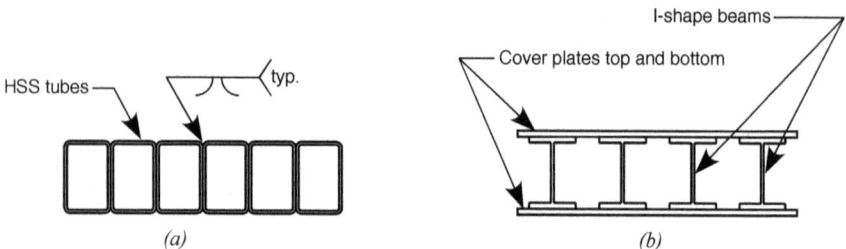

Figure 3.11 Steel Crawler Crane Mats

A steel mat for use under an outrigger will generally be more complex in its configuration. Fig. 3.12 illustrates a style of outrigger mat used by the author for a wide range of outrigger-supported cranes. The beam grillage layout provides effective load spreading in both directions, thus allowing both dimensions of the mat to be selected to provide the best fit with respect to the clearances around the crane and for transportation of the mats. The size and number of beams in each direction, along with the thicknesses of the top and bottom cover plates, are selected to optimize the balance of strength and weight. The overall depth of the mat should also be limited to something a little less than the ground clearance of the retracted outriggers of the cranes with which the mats will be used so that the crane can be driven into position over the outrigger mats.

The structural analysis of a steel crane mat is quite straightforward, since only simple bending, shear, concentrated loads to the beam webs, and welding of the components need to be addressed. Engineers in the U.S. typically use the American Institute of Steel Construction *Specification for Structural Steel Buildings* (AISC 1989 or AISC 2016) for evaluating the mat strength. Both of these versions of the specification provide a nominal design factor of about 1.67 with respect to the applicable limit states (i.e., yielding, local buckling, etc.). Other comparable design standards may also be used for steel mat design.

It is noted that there are no construction industry standards or regulations that address the design or fabrication of steel crane mats, so the responsible engineer must decide on the appropriate methods and design factors to be applied. Additional comments about the design of fabricated steel crane mats are offered in Section 4.3.1.

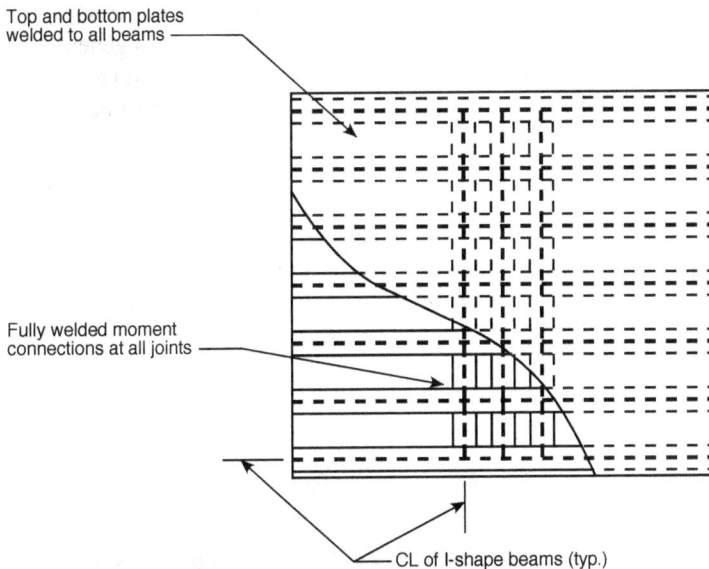

Figure 3.12 Steel Grillage Outrigger Mat

3.7   DESIGN VALUES FOR OTHER MATERIALS

Crane mats, most commonly for use under outriggers, are now being produced by a handful of vendors using specialty materials. These include laminated wood products and different type of synthetic materials, such as ultra high molecular weight (UHMW) plastic and fiber-reinforced polymer (FRP). These products offer a number of advantages, but also have limitations.

The laminated wood mat products include very thick plywood outrigger mats and conventional looking mats that have "timbers" that are built up from boards pressure-glued together, such as emtek® mats. The primary advantage offered by these products is greater strength. By starting with thinner boards, the individual pieces can be selected from higher grades of wood (that is, pieces with fewer or smaller defects), thus offering higher allowable stresses. The full advantage of these higher allowable stresses is not always realized, though. As is discussed in Chapter 4, the load that a long crane mat can safely support is limited by deflection as well as by strength. Thus, the usable capacity of some higher strength wood products may be limited by their flexibility.

The synthetic materials currently used for crane mat products all have one significant advantage: They don't rot. The various plastics used today are completely impervious to the effects of moisture, termites, fungi, and the other natural enemies of wood. Like the steel grillage mats, the synthetic outrigger mats can be designed to be fully effective in spreading the applied load in both directions and they can be manufactured in various shapes and sizes. There are limitations to their use, however. First, due to the limits of the manufacturing processes, the maximum size synthetic mat that can be produced is not particularly large. Thus, these products are most applicable to small to medium capacity cranes. Second, since synthetic mats are proprietary products, the mat user will generally not have the raw data, such as material properties, to allow independent calculation of the mat strength and stiffness. Rather, a rated load for the mat must be obtained from the mat manufacturer. From a practical viewpoint, though, this is very rarely a problem with the types of cranes and types of jobs for which these synthetic mats are a good fit.

The last material to be considered here is aluminum. Just as structural steel beams and plates can be used to fabricate mats, aluminum shapes and plates can also be used. The *Specification for Aluminum Structures* AA (2015) is a building design standard for aluminum structures that is comparable to the AISC specifications for the design of steel structures. This is a useful standard that may be applied to the design of aluminum crane mats.

3.8   PLYWOOD USED UNDER CRANES

Plywood is often used to pack out gaps between crane mats and to level crawler cranes. In these locations, the plywood sheets are loaded in compression perpendicular to the surfaces. As such, specific calculations to evaluate the performance of plywood shims are typically not required. The following information is provided to support calculations when needed.

The basic allowable compression stress $F_{c\perp}$ of standard plywood sheets is given in APA (2013) as 360 psi in dry service, defined as the condition in which the wood has a moisture content of 16% or less. This value is based on a 95% confidence interval on the average bearing stress at a deformation of 0.04 inch (1.0 mm) and an adjustment factor of 1.67. The data underlying this allowable stress come from the testing of hundreds of specimens, as reported in APA (2013). The allowable compression stress is reduced by 50% for wet service. These allowable stresses are shown in different units in Table 3.15.

3.9 APPLICATION SUMMARY

This chapter contains detailed information, including the results of two studies, by which the strength of a crane mat can be calculated. This closing section of the chapter summarizes how this material should be used in the design of mobile crane support.

The strength of a timber crane mat is most commonly evaluated using an allowable stress (or strength) design approach. Allowable stresses for the species of timbers commonly used in the United States for crane mat construction as well as a selection of tropical hardwood species globally used for mat construction are given in Table 3.8. The derivation of these allowable stresses accounts for the differences between actual timber dimensions and the nominal dimensions and for common defects such as knots and wane, as determined by the dimension and condition study reported in this chapter. Thus, timber mat strength calculations can be performed using the nominal timber dimensions in conjunction with the allowable stresses given in Table 3.8.

If mats made of wood species not addressed in this chapter are to be used, suitable design properties can be calculated using Eq. 3.3 for bending, Eq. 3.4 for shear, Eq. 3.5 for compression perpendicular to the grain, and Eq. 3.6 for the modulus of elasticity. The use of these equations requires clear wood stress values for the species of interest from a reliable source.

Fabricated steel crane mats are most commonly designed in the U.S. using the AISC *Specification for Structural Steel Buildings* (either the 1989 or 2016 edition). Both editions of this specification provide a nominal design factor of

TABLE 3.15 Allowable Compression Stresses $F_{c\perp}$ for Plywood

Condition	psi	psf	kPa
Dry Service	360	51,800	2,480
Wet Service	180	25,900	1,240

1.67. Comparable steel design standards in use in other countries may be applied to mat construction, if deemed suitable by the responsible lift planner/engineer.

The performance characteristics of crane mats made of proprietary materials typically must be obtained from the mat manufacturer. Understandably, the details of these products are often treated as trade secrets.

3.10 ACKNOWLEDGEMENTS

The following companies generously supported the crane mat dimensions and conditions study discussed in Section 3.1 by providing measurements, photographs, and descriptions of timber crane mats. Their assistance is gratefully acknowledged.

American Mat & Timber Co., Inc., Houston, TX
Barnhart Crane & Rigging Co., Memphis, TN
Bragg Crane & Rigging Co., Richmond, CA
Burkhalter Rigging, Inc., Columbus, MS
Crane Rental Corporation, Davenport, FL
Deep South Crane & Rigging Co., Baton Rouge, LA
EarthFax Engineering, Inc., Salt Lake City, UT
Lampson International, Inc., Denver, CO; Kennewick, WA
Maxim Crane Works, Benica, CA; Long Beach, CA; Wilder, KY
Mr. Crane, Inc., Orange, CA
Southern Crane, Raleigh, NC
Taylor Crane & Rigging, Inc., Coffeyville, KS
W.J. Casey Trucking & Rigging Co., Inc, Branchburg, NJ
White Construction, Inc., Robinson, IL.

3.11 REFERENCES

Aluminum Association (AA) (2015), *Specification for Aluminum Structures*, Arlington, VA.

American Institute of Steel Construction (AISC) (1989), *Specification for Structural Steel Buildings – Allowable Stress Design and Plastic Design*, Chicago, IL.

American Institute of Steel Construction (AISC) (2016), *Specification for Structural Steel Buildings*, Chicago, IL.

American Wood Council (AWC) (2016), ANSI/AWC NDS-2015 *National Design Specification® for Wood Construction* (NDS), Leesburg, VA.

APA – The Engineered Wood Association (2013), TT-001B *Allowable Bearing Stress for APA Wood Structural Panels*, Tacoma, WA.

ASTM International (2011), D245-06(2011) *Standard Practice for Establishing Structural Grades and Related Allowable Properties for Visually Graded Lumber*, West Conshohocken, PA.

ASTM International (2012), D9-12 *Standard Terminology Relating to Wood and Wood-Based Products*, West Conshohocken, PA.

ASTM International (2016), SI 10-2016 *American National Standard for Metric Practice*, West Conshohocken, PA.

ASTM International (2017), D2555-17a *Standard Practice for Establishing Clear Wood Strength Values*, West Conshohocken, PA.

British Standards Institution (BSI) (2016), BS EN 16737 *Structural Timber – Visual Strength Grading of Tropical Hardwood*, London, U.K.

Forest Products Laboratory (FPL) (1984), Agriculture Handbook 607 *Tropical Timbers of the World*, Madison, WI.

Forest Products Laboratory (FPL) (2010), General Technical Report FPL-GTR-190 *Wood Handbook – Wood as an Engineering Material*, Madison, WI.

Guyana Forestry Commission (GFC) (2002), *Guyana Timber Grading Rules for Hardwoods*, Georgetown, Guyana.

National Forest Products Association (NFPA) (1986), *National Design Specification for Wood Construction* (NDS), Washington, D.C.

National Lumber Grades Authority (NLGA) (2014), *Standard Grading Rules for Canadian Lumber*, New Westminster, BC.

Northeastern Lumber Manufacturers Association, Inc. (NeLMA) (2013), *Standard Grading Rules for Northeastern Lumber*, Cumberland Center, ME.

Siegrist, C.A. (2002), "Flexural Testing of Timber Mats," *Report FL200.02.1157.1*, ATLSS Engineering Research Center, Lehigh University, Bethlehem, PA.

West Coast Lumber Inspection Bureau (WCLIB) (2015), Standard No. 17 *Grading Rules for West Coast Lumber*, Tigard, OR.

4 Crane Mat Behavior

One very important aspect of evaluating the performance of a crane mat is the analysis of the interaction between the mat and the soil. A long crane mat will not necessarily bear on the soil over its entire length. The interaction of bending of the mat and compression of the soil will alter the actual length of bearing of the mat. The analysis of the behavior of a crane mat requires a determination of the length of the mat that actually bears on the soil and contributes to the support of the crane. At working loads, this is a relatively simple "beam on an elastic foundation" problem. However, such a solution may not produce a realistic result as the ultimate bearing capacity is approached due to the nonlinearity of the soil. Further, the elastic properties of the soil needed to perform such an analysis are often not available. Thus, while engineers generally have the tools necessary to perform these calculations, seeking a theoretically "exact" crane mat analysis is not a practical goal due to this lack of reliable input values.

In this chapter, a practical means is developed by which the effective bearing length of a crane mat can be calculated. This method is based on readily available values and produces an acceptably safe and reliable design result. Although validated by nonlinear finite element analyses, the method developed here is fundamentally based on crane mat design practices that have been in use for many years. Brought together with the crane load calculations, ground bearing capacity, and mat strength information developed in Chapters 1, 2 and 3, respectively, a job-ready crane mat design method is developed.

Some readers will recognize the material in Sections 4.1 and 4.2. The original development of this crane mat design method was first published in a paper by the author titled "Effective Bearing Length of Crane Mats" (Duerr 2010). This paper was distributed in conjunction with a presentation titled "Crane Mats and Ground Bearing Issues" given at the 2010 Crane & Rigging Conference in Houston. The discussion here is expanded for improved clarity, to cover additional aspects of mat design, and to tie into the material covered in Chapters 1, 2, and 3.

4.1 PAST PRACTICE OF CRANE MAT ANALYSIS

Lift planners and lift specialist engineers in construction for many years have used a number of different approaches to size crane mats. The two most common of these methods are described in this section. These design methods, both relatively simple in their approach, form the basis for the crane mat design procedure developed in this chapter.

4.1.1 Mat Length Based on Ground Bearing Capacity

This crane mat design method is the simplest and most straightforward. Once the load from the crane has been calculated as covered in Chapter 1, whether an outrigger load or a crawler track pressure, the required crane mat area is calculated by dividing the crane load plus the weight of the mat by the allowable ground bearing pressure. Dividing this area by the width of the mat gives us the required effective bearing length of the mat. This mat effective bearing length is then used to calculate the bending and shear stresses in the mat, based on the assumption of a uniform pressure equal to the crane load divided by the bearing area acting upward on the bottom of the mat. If the calculated stresses in the mat are equal to or less than the corresponding allowable stresses, the mat is acceptable for this particular load and ground bearing capacity.

Fig. 4.1 is an illustration of this basic crane mat arrangement. Shown are the crane mat, the crane load shown as applied through an outrigger, the bearing length of the mat on the soil, and the assumed uniform bearing pressure between the mat and the soil. This crane mat design method can be written in equation form as Eqs. 4.1 through 4.8.

Timber design practice (AWC 2016) allows the evaluation of the shear force in a beam subject to a uniformly distributed load at a point located a distance from the face of the support equal to the depth of the beam. The shear force in steel or aluminum beams is calculated at the face of the support. Eqs. 4.7a and 4.8a are written for the design of timber crane mats. Eqs. 4.7b and 4.8b are written for the design of steel or aluminum mats. The appropriate equations for mats made of other structural materials must be determined based on the applicable design practices for those materials.

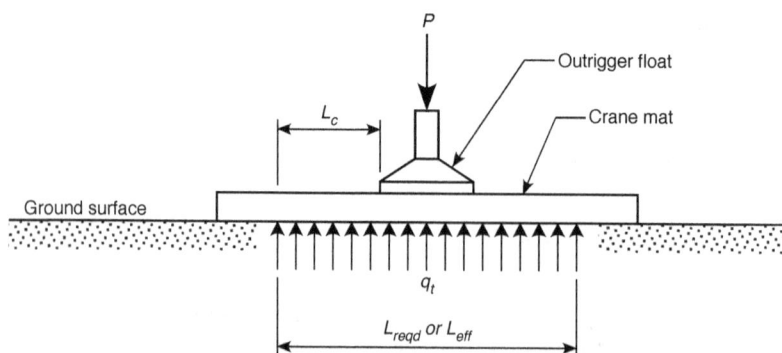

Figure 4.1 Simple Crane Mat Arrangement

$$A_{reqd} = \frac{P+W}{q_a} \tag{4.1}$$

$$L_{reqd} = \frac{A_{reqd}}{B} \tag{4.2}$$

$$L_c = \frac{L_{reqd} - C}{2} \tag{4.3}$$

$$q = \frac{P}{L_{reqd}B} \tag{4.4}$$

$$M = \frac{(qB)L_c^2}{2} \tag{4.5}$$

$$f_b = \frac{M}{S_x} \le F_b \tag{4.6}$$

$$V = (qB)(L_c - d) \tag{4.7a}$$

$$V = (qB)L_c \tag{4.7b}$$

$$f_v = \frac{1.5V}{Bd} \le F_v \tag{4.8a}$$

$$f_v = \frac{V}{n_w d t_w} \le F_v \tag{4.8b}$$

where

P	=	crane load applied to one mat;
W	=	self-weight of the crane mat;
q_a	=	allowable ground bearing pressure;
A_{reqd}	=	required crane mat bearing area;
B	=	crane mat width;
L_{reqd}	=	required effective bearing length of the crane mat;
C	=	bearing width of the track or outrigger float;
L_c	=	cantilevered length of the crane mat;
q	=	ground bearing pressure due to the crane load P;
M	=	bending moment in the crane mat;
S_x	=	section modulus of the crane mat;
f_b	=	calculated bending stress due to M;
F_b	=	allowable bending stress;
V	=	shear in the crane mat;

d =   crane mat depth (or thickness);
f_v =   calculated shear stress due to V;
F_v =   allowable shear stress;
n_w =   number of webs (steel or aluminum beam design); and,
t_w =   web thickness.

The allowable stresses for timber mats to be used in these calculations are as developed in Section 3.5. The allowable stresses for fabricated steel mats should be based on suitable structural design methods, as discussed briefly in Section 3.6. Determination of the appropriate allowable stresses to be used for other materials, particularly for proprietary synthetics, generally requires working with the mat vendor, as discussed in Section 3.7.

Application of this mat design approach based on the ground bearing capacity is illustrated in Example 4-1.

EXAMPLE 4-1

Analyze the mat arrangement shown in Fig. 4.1 for an outrigger load of 100,000 pounds applied to a 12" x 4' x 20' timber crane mat through an outrigger float that is 24" wide along the length of the timbers (assume that the float bears on all timbers). The allowable ground bearing pressure is 3,000 psf and mat allowable stresses, as developed in Section 3.5, are F_b = 1,400 psi and F_v = 200 psi. The timber density is taken as 50 pcf.

Mat dead weight W = 1' x 4' x 20' x 50 pcf = 4,000 pounds
The mat is now analyzed using Eqs. 4.1 through 4.6, 4.7a and 4.8a:

$$A_{reqd} = \frac{P+W}{q_a} = \frac{100,000+4,000}{3,000} = 34.67 \text{ ft}^2$$

$$L_{reqd} = \frac{A_{reqd}}{B} = \frac{34.67}{4.00} = 8.67 \text{ feet}$$

$$L_c = \frac{L_{reqd}-C}{2} = \frac{8.67-24/12}{2} = 3.33 \text{ feet}$$

$$q = \frac{P}{L_{reqd}B} = \frac{100,000}{8.67 \times 4.00} = 2,885 \text{ psf}$$

$$M = \frac{(qB)L_c^2}{2} = \frac{(2,885 \times 4.00)3.33^2}{2} = 64,103 \text{ pound-feet} = 769,231 \text{ pound-inches}$$

$$f_b = \frac{M}{S_x} = \frac{769,231}{12^2(48)/6} = 668 \text{ psi} < F_b = 1,400 \text{ psi}$$

$$V = (qB)(L_c-d) = (2,885 \times 4.00)(3.33-1.00) = 26,923 \text{ pounds}$$

$$f_v = \frac{1.5V}{Bd} = \frac{1.5 \times 26,923}{48.00 \times 12.00} = 70 \text{ psi} < F_v = 200 \text{ psi}$$

Since both the calculated bending stress and calculated shear stress are less than their respective allowable stresses, the mat is shown to be acceptable for this application, based on the ground surface being loaded to its allowable bearing pressure.

4.1.2   Mat Length Based on Mat Strength

This design approach is the reverse of the method described in Section 4.1.1. Here, the effective bearing length L_{eff} of the crane mat is assumed, the ground bearing pressure and mat stresses calculated, and then the assumed bearing length is adjusted and the calculations repeated until the resulting bending stress or shear stress reaches its corresponding allowable stress. The ground bearing pressure is then computed using this effective bearing length, again based on the assumption of uniform bearing between the mat and the soil. If the calculated bearing pressure is equal to or less than the allowable bearing pressure, the mat is acceptable.

Again, we can write the design method in equation form. As was done for the first method, Eqs. 4.13a and 4.14a are written for the design of timber crane mats and Eqs. 4.13b and 4.14b are written for the design of steel or aluminum mats.

$$L_c = \frac{L_{eff} - C}{2} \tag{4.9}$$

$$q = \frac{P}{L_{eff} B} \tag{4.10}$$

$$M = \frac{(qB) L_c^2}{2} \tag{4.11}$$

$$f_b = \frac{M}{S_x} = F_b \tag{4.12}$$

$$V = (qB)(L_c - d) \tag{4.13a}$$

$$V = (qB) L_c \tag{4.13b}$$

$$f_v = \frac{1.5V}{Bd} = F_v \tag{4.14a}$$

$$f_v = \frac{V}{n_w d t_w} = F_v \tag{4.14b}$$

$$q_t = \frac{P + W}{L_{eff} B} \leq q_a \tag{4.15}$$

where

L_{eff} =   effective bearing length of the crane mat;

q_t =   calculated total ground bearing pressure; and,

all other terms are as previously defined.

Note that this second method is iterative. A value of L_{eff} must be assumed, Eqs. 4.9 through 4.14 solved, and then L_{eff} adjusted as necessary to satisfy the equalities of Eqs. 4.12 and 4.14. The design is complete (that is, the effective bearing length has been determined) when either the bending stress or the shear stress reaches its allowable value. The value of L_{eff} calculated using Eqs. 4.9 through 4.14 is then used in Eq. 4.15 to determine the total ground bearing pressure q_t acting on the soil. If this value of q_t is less than or equal to the allowable ground bearing pressure, the mat is acceptable for this particular load.

When the crane mat is made of the most commonly used wood species and the allowable ground bearing pressure is not unusually high, bending usually governs the mat design. However, both shear and bending must always be checked.

As with the ground bearing capacity method, we can illustrate the application of the mat strength method by means of an example.

EXAMPLE 4-2

Analyze the mat arrangement shown in Fig. 4.1 for an outrigger load of 100,000 pounds applied to a 12" x 4' x 20' timber crane mat through an outrigger float that is 24" wide along the length of the timbers (assume that the float bears on all timbers). The allowable ground bearing pressure is 3,000 psf and mat allowable stresses, as developed in Section 3.5, are F_b = 1,400 psi and F_v = 200 psi. The timber density is taken as 50 pcf.

Mat dead weight W = 1' x 4' x 20' x 50 pcf = 4,000 pounds

The mat is now analyzed using Eqs. 4.9 through 4.12, 4.13a, 4.14a, and 4.15:

Assume L_{eff} = 14.48 feet

$$L_c = \frac{L_{eff} - C}{2} = \frac{14.48 - 24/12}{2} = 6.24 \text{ feet}$$

$$q = \frac{P}{L_{eff}B} = \frac{100,000}{14.48 \times 4.00} = 1,727 \text{ psf}$$

$$M = \frac{(qB)L_c^2}{2} = \frac{(1,727 \times 4.00)6.24^2}{2} = 134,400 \text{ pound-feet} = 1,612,800 \text{ pound-inches}$$

$$f_b = \frac{M}{S_x} = \frac{1,612,800}{12^2 \times 48/6} = 1,400 \text{ psi} = F_b = 1,400 \text{ psi}$$

$$V = (qB)(L_c - d) = (1,727 \times 4.00)(6.24 - 1.00) = 36,184 \text{ pounds}$$

$$f_v = \frac{1.5V}{Bd} = \frac{1.5 \times 36,184}{48.00 \times 12.00} = 94 \text{ psi} < F_v = 200 \text{ psi}$$

$$q_t = \frac{P+W}{L_{eff}B} = \frac{100,000+4,000}{14.48 \times 4.00} = 1,796 \text{ psf} < q_a = 3,000 \text{ psf}$$

The assumed value of L_{eff} is shown to produce a bending stress equal to the allowable bending stress, a shear stress less than the allowable shear stress, and a ground bearing pressure less than the allowable pressure. Thus, the mat is shown to be acceptable for this application (as was found in Example 4-1).

4.1.3   Comments on These Design Methods

Both of these crane mat design methods are in popular use in the United States and generally give adequate results. However, there is one important shortcoming in the way these calculations are commonly applied in practice. Neither method shows how close a particular design is to reaching its allowable load carrying limit. The first method loads the ground surface to its allowable bearing capacity and then shows that the stresses in the mats are equal to or less than their allowable values. The second method loads the mats to the allowable bending or shear capacity and then shows that the ground bearing pressure is equal to or less than the allowable bearing pressure.

We can examine this problem by comparing the results found in Examples 4-1 and 4-2. The calculation results from these two examples using an applied load of 100,000 pounds (444,822 newtons) are summarized side by side in Table 4.1 in U.S. customary units (USCU) and SI equivalents. Both methods show that the mat design is acceptable, but the design margin is not obvious. The Ground Bearing Capacity method shows that the applied ground bearing pressure is equal to the allowable bearing pressure, the mat bending stress is 668 psi, or 48% of the allowable bending stress, and the mat shear stress is 70 psi, or 35% of the allowable shear stress. The Mat Strength method shows that the mat is loaded to 100% of its allowable bending stress and 47% of the allowable shear stress and that the ground bearing pressure is 1,796 psf, or 60% of the allowable bearing capacity. Not clear is the load carrying margin that remains for this mat/soil combination. That is, we can't see the true utilization ratio that is provided.

This crane mat analysis is now repeated using a load of 135,256 pounds (601,649 newtons) and both crane mat design methods. All other values remain the same. The results of this analysis are summarized in Table 4.2. (Although not shown here, the calculations performed to develop the values in Table 4.2 are identical to those used in Examples 4-1 and 4-2.)

Here we see that the two mat design methods converge at the load where both the mat strength and the ground bearing capacity limits are reached. This example shows us that the capacity of this mat on this soil is a crane load of 135,256 pounds and that the mat capacity is limited by its bending strength. Thus, the crane load of 100,000 pounds used in Examples 4-1 and 4-2 loaded the mat/soil combination to 74% of its capacity.

TABLE 4.1 Design Method Comparison – Examples 4-1 and 4-2

	Ground Bearing Cap. Method		Mat Strength Method	
	USCU	SI	USCU	SI
Mat Weight, W	4,000 lbs	17,793 N	4,000 lbs	17,793 N
A_{reqd} (Eq. 4.1)	34.67 ft^2	3.22 m^2		
L_{reqd} (Eq. 4.2)	8.67 feet	2.64 meters		
L_c (Eq. 4.3)	3.33 feet	1.02 meters		
Assumed L_{eff}			14.48 feet	4.41 meters
L_c (Eq. 4.9)			6.24 feet	1.90 meters
q (Eq. 4.4; Eq. 4.10)	2,885 psf	138,116 Pa	1,727 psf	82,691 Pa
M (Eq. 4.5; Eq. 4.11)	769,231 lb-in	86,912 N-m	1,612,800 lb-in	182,222 N-m
f_b (Eq. 4.6; Eq. 4.12)	668 psi	4,604 kPa	1,400 psi	9,653 kPa
V (Eq. 4.7a; Eq. 4.13a)	26,923 lbs	119,760 N	36,184 lbs	160,953 N
f_v (Eq. 4.8a; Eq. 4.14a)	70 psi	483 kPa	94 psi	650 kPa
q_t (Eq. 4.15)	3,000 psf	143,641 Pa	1,796 psf	85,999 Pa

This margin of 74% of capacity found in this example cannot be seen in the calculations detailed in Examples 4-1 and 4-2 and summarized in Table 4.1. Although these commonly used crane mat design methods generally provide results that are acceptably safe, they do not provide an indication of the percent utilization (or demand/capacity ratio) of the mat/soil combination. As such an expression of calculation results is often desirable in order to demonstrate the margin provided by a proposed crane support configuration, the concepts of these past practice methods are used in Section 4.2 to develop a design method that provides this result.

TABLE 4.2 Design Method Comparison – P = 135,256 lbs. (601,649 N)

	Ground Bearing Cap. Method		Mat Strength Method	
	USCU	SI	USCU	SI
Mat Weight, W	4,000 lbs	17,793 N	4,000 lbs	17,793 N
A_{reqd} (Eq. 4.1)	46.42 ft^2	4.31 m^2		
L_{reqd} (Eq. 4.2)	11.60 feet	3.54 meters		
L_c (Eq. 4.3)	4.80 feet	1.46 meters		
Assumed L_{eff}			11.60 feet	3.54 meters
L_c (Eq. 4.9)			4.80 feet	1.46 meters
q (Eq. 4.4; Eq. 4.10)	2,914 psf	139,515 Pa	2,914 psf	139,515 Pa
M (Eq. 4.5; Eq. 4.11)	1,612,800 lb-in	182,222 N-m	1,612,800 lb-in	182,222 N-m
f_b (Eq. 4.6; Eq. 4.12)	1,400 psi	9,653 kPa	1,400 psi	9,653 kPa
V (Eq. 4.7a; Eq. 4.13a)	44,317 lbs	197,134 N	44,317 lbs	197,134 N
f_v (Eq. 4.8a; Eq. 4.14a)	115 psi	796 kPa	115 psi	796 kPa
q_t (Eq. 4.15)	3,000 psf	143,641 Pa	3,000 psf	143,641 Pa

Based on feedback from the Duerr (2010) paper, it has become obvious that one important clarification is needed here. These descriptions and examples of the earlier calculation methods are shown for reference and to provide the basis for the balanced design method developed in the next section. They are not recommended for use.

4.2   BALANCED MAT ANALYSIS METHOD

A practical method of crane mat design can be derived that is based on the accepted past practice described in Section 4.1. Unlike the calculations discussed in Section 4.1, however, this approach shows the utilization of the mat strength and the ground bearing capacity. This method uses as input only values that are routinely available and is based on balancing the mat strength and stiffness with the ground bearing capacity, expressed as the allowable ground bearing pressure.

4.2.1   Effective Bearing Area Calculation Method

Consider first the bending strength of the mat. We wish to determine the effective bearing length L_{eff} at which both the allowable bending strength of the mat and the allowable ground bearing pressure are reached simultaneously. This can be done by expressing q in terms of q_a (Eq. 4.16) and then writing Eq. 4.11 in terms of this expression for q, Eq. 4.9, and the allowable moment of the mat M_n, which results in Eq. 4.17. By rearranging terms, Eq. 4.17 can be written as Eq. 4.18. The last term of this equation can be shown to be trivial, so Eq. 4.18 reduces to Eq. 4.19, which is a quadratic equation in which the quantity in the first set of parentheses is a, the quantity in the second set is b, and the quantity in the third set is c. The standard solution of a quadratic equation is shown in Eq. 4.20.

One remaining question here is: What is the appropriate value to use for the allowable moment M_n of a crane mat? When considering a common timber crane mat, the allowable moment is simply the section modulus of the mat based on the nominal dimensions multiplied by the appropriate allowable bending stress, as summarized in Table 3.8 (Eq. 4.21). When designing a mat to be fabricated from steel, aluminum, or some other structural material, a suitable design standard must be applied. Even today, the old allowable stress design AISC specification (AISC 1989) is still used by many engineers for the design of lifting equipment not covered by other more specific standards. Other building design standards, such as AISC (2016) and AA (2015) may also be applied to crane mat design. Evaluation of a mat made of a proprietary material will require input from the mat manufacturer.

$$q = q_a - \frac{W}{L_{eff} B} \tag{4.16}$$

$$M_n = \frac{\left(q_a - \frac{W}{L_{eff}B}\right)B\left(\frac{L_{eff} - C}{2}\right)^2}{2} \tag{4.17}$$

$$\left(q_a B\right)L_{eff}^2 + \left(-2q_a BC - W\right)L_{eff} + \left(q_a BC^2 + 2CW - 8M_n\right) - \frac{C^2 W}{L_{eff}} = 0 \tag{4.18}$$

$$\left(q_a B\right)L_{eff}^2 + \left(-2q_a BC - W\right)L_{eff} + \left(q_a BC^2 + 2CW - 8M_n\right) = 0 \tag{4.19}$$

$$L_{eff} = \frac{-b \pm \sqrt{b^2 - 4ac}}{2a} \le L_{mat} \tag{4.20}$$

$$M_n = F_b S_x \tag{4.21}$$

where L_{mat} is the actual length of the crane mat.

A solution of L_{eff} must also be made based on the shear strength of the crane mat in which the allowable shear stress and allowable ground bearing pressure are reached simultaneously. Eqs. 4.22 through 4.24 are based on timber design practice. Here, Eq. 4.13a is written in terms of Eqs. 4.9 and 4.16 and the allowable shear strength of the mat V_n, which results in Eq. 4.22. This equation is then rewritten in quadratic form as Eq. 4.23 where the quantity in the first set of parentheses is a, the quantity in the second set is b, and the quantity in the third set is c. Eq. 4.20 is again used to solve for L_{eff}. The allowable shear strength V_n is given by Eq. 4.24.

$$V_n = \left(q_a - \frac{W}{L_{eff}B}\right)B\left(\frac{L_{eff} - C}{2} - d\right) \tag{4.22}$$

$$\left(q_a B\right)L_{eff}^2 + \left(-2V_n - q_a BC - 2q_a Bd - W\right)L_{eff} + \left(WC + 2Wd\right) = 0 \tag{4.23}$$

$$V_n = F_v \frac{Bd}{1.5} \tag{4.24}$$

When designing mats made of steel or aluminum, the shear strength is evaluated at the point of maximum shear, rather than at a distance d from that point. Thus, Eqs. 4.13b, 4.22, and 4.23 are rewritten for steel or aluminum design as Eqs. 4.25, 4.26, and 4.27. The allowable shear strength is given by Eq. 4.28.

$$V_n = \left(qB\right)L_c \tag{4.25}$$

$$V_n = \left(q_a - \frac{W}{L_{eff}B}\right)B\left(\frac{L_{eff}-C}{2}\right) \qquad (4.26)$$

$$\left(q_aB\right)L_{eff}^2 + \left(-2V_n - q_aBC - W\right)L_{eff} + WC = 0 \qquad (4.27)$$

$$V_n = F_v n_w dt_w \qquad (4.28)$$

Last, a limit of the effective bearing length based on the deflection of the mat is proposed. Examination of numerous mat design examples using only the criteria of bending and shear strength shows that some mats exhibit excessive deflections [greater than one inch (25 mm)] on softer soils. Recognizing that the mat/soil combination must provide firm support to the crane, excessive deflection of the mats and soil is not acceptable. Therefore, we should limit the effective bearing length based on the stiffness of the mats. This is a more difficult criterion to define, since there isn't a well established deflection limit state as exist for bending and shear. A design deflection limit of 0.75% of L_c is suggested. This limit is based on an examination of numerous past mat designs that performed acceptably.

The deflection of a crane mat is commonly calculated by treating the mat as a cantilever beam of length L_c and loaded by an upward uniform pressure equal to q. We can express this in equation form as Eq. 4.29.

$$\Delta = \frac{\left(qB\right)L_c^4}{8EI_x} \qquad (4.29)$$

where
Δ =   vertical deflection;
E =   modulus of elasticity;
I_x =   moment of inertia of the mat; and,
            all other terms are as previously defined.

The modulus of elasticity values tabulated in Chapter 3 and which have been developed from data in AWC (2016) are adjusted values that account for shear deflection. Therefore, a separate calculation of shear deflection is not necessary.

This deflection criterion typically will only control the effective bearing length with softer soils. Examination of such designs shows us that $q \approx 0.9\, q_a$. If we let $\Delta = 0.0075\, L_c$ and use this approximation for q, we can easily solve Eq. 4.29 for L_c (Eq. 4.30) and L_{eff}(Eq. 4.31).

$$L_c = \sqrt[3]{\frac{0.06EI_x}{0.9q_aB}} \qquad (4.30)$$

$$L_{eff} = 2L_c + C \le L_{mat} \qquad (4.31)$$

With these three sets of calculations complete, the smallest value of L_{eff} based on the moment and shear strength analyses and the deflection analysis is taken as the effective bearing length of the crane mat. The mat and the soil are then evaluated based on the usual assumption of a uniform bearing pressure q between the mat and the soil over the effective bearing area using Eqs. 4.32 through 4.36. The shear strength calculations are done using Eq. 4.35a for timber mats and Eq. 4.35b for steel or aluminum mats. As noted before, mats of other structural materials must be checked using appropriate methods.

$$q = \frac{P}{L_{eff} B} \tag{4.32}$$

$$L_c = \frac{L_{eff} - C}{2} \tag{4.33}$$

$$M = \frac{(qB) L_c^2}{2} \le M_n \tag{4.34}$$

$$V = qB(L_c - d) \le V_n \tag{4.35a}$$

$$V = qBL_c \le V_n \tag{4.35b}$$

$$q_t = \frac{P + W}{L_{eff} B} \le q_a \tag{4.36}$$

Following conventional engineering practice, the results of the design are best expressed as utilization ratios for each limit state (mat bending, mat shear, ground bearing pressure), where the utilization ratio is the actual force, stress, or pressure divided by the corresponding allowable value. The ability to calculate a utilization ratio is a particular advantage of this method.

Application of this calculation method is illustrated in Example 4-3 using the same load and mat configuration as was used in Examples 4-1 and 4-2.

EXAMPLE 4-3

Analyze the mat arrangement shown in Fig. 4.1 for an outrigger load of 100,000 pounds applied to a 12" x 4' x 20' timber crane mat through an outrigger float that is 24" wide along the length of the timbers (assume that the float bears on all timbers). The allowable ground bearing pressure is 3,000 psf and mat allowable stresses, as developed in Section 3.5, are F_b = 1,400 psi and F_v = 200 psi. The timber density is taken as 50 pcf.

Mat dead weight W = 1' x 4' x 20' x 50 pcf = 4,000 pounds

The effective bearing length of the mat is now calculated using Eqs. 4.19 through 4.31, as applicable:

Calculate the effective bearing length L_{eff} as limited by bending strength.

$$M_n = F_b S_x = 1,400 \frac{12^2 48}{6} = 1,612,800 \text{ pound-inches} = 134,400 \text{ pound-feet}$$

$$\left(q_a B\right) L_{eff}^2 + \left(-2 q_a BC - W\right) L_{eff} + \left(q_a BC^2 + 2CW - 8M_n\right) = 0$$

$$\left(3,000 \times 4.00\right) L_{eff}^2 + \left(-2 \times 3,000 \times 4.00 \times 2.00 - 4,000\right) L_{eff}$$

$$+ \left(3,000 \times 4.00 \times 2.00^2 + 2 \times 2.00 \times 4,000 - 8 \times 134,400\right) = 0$$

$$\left(12,000\right) L_{eff}^2 + \left(-52,000\right) L_{eff} + \left(-1,011,200\right) = 0$$

$$L_{eff} = \frac{-\left(-52,000\right) \pm \sqrt{\left(-52,000\right)^2 - 4\left(12,000\right)\left(-1,011,200\right)}}{2\left(12,000\right)} = 11.599 < L_{mat} = 20.000$$

Calculate the effective bearing length L_{eff} as limited by shear strength.

$$V_n = F_v \frac{Bd}{1.5} = 200 \frac{48.00 \times 12.00}{1.5} = 76,800 \text{ pounds}$$

$$\left(q_a B\right) L_{eff}^2 + \left(-2V_n - q_a BC - 2q_a Bd - W\right) L_{eff} + \left(WC + 2Wd\right) = 0$$

$$\left(3,000 \times 4.00\right) L_{eff}^2$$

$$+ \left(-2 \times 76,800 - 3,000 \times 4.00 \times 2.00 - 2 \times 3,000 \times 4.00 \times 1.00 - 4,000\right) L_{eff}$$

$$+ \left(4,000 \times 2.00 + 2 \times 4,000 \times 1.00\right) = 0$$

$$\left(12,000\right) L_{eff}^2 + \left(-205,600\right) L_{eff} + \left(16,000\right) = 0$$

$$L_{eff} = \frac{-\left(-205,600\right) \pm \sqrt{\left(-205,600\right)^2 - 4\left(12,000\right)\left(16,000\right)}}{2\left(12,000\right)} = 17.055 < L_{mat} = 20.000$$

Calculate the effective bearing length L_{eff} as limited by deflection.

$$L_c = 3\sqrt{\frac{0.06 E I_x}{0.9 q_a B}} = 3\sqrt{\frac{0.06\left(1,200,000\right) 12.00^3 48.00/12}{0.9\left(3,000/144\right) 48.000}} = 82.079 \text{ inches} = 6.840 \text{ feet}$$

$$L_{eff} = 2L_c + C = 2 \times 6.840 + 2.00 = 15.680 < L_{mat} = 20.000$$

The smallest value of L_{eff} calculated above is 11.599 feet, as limited by bending strength. Thus, this value is the effective bearing length of the mat for this application The mat forces and ground bearing pressure are now calculated as follows, using Eqs. 4.32 through 4.36.

The calculated bending moment, shear, and ground bearing pressure are expressed relative to the allowable values as utilization ratios.

$$q = \frac{P}{L_{eff}B} = \frac{100,000}{11.599 \times 4.00} = 2,155 \text{ psf}$$

$$L_c = \frac{L_{eff} - C}{2} = \frac{11.599 - 24/12}{2} = 4.799 \text{ feet}$$

$$M = \frac{(qB)L_c^2}{2} = \frac{(2,155 \times 4.00)4.799^2}{2} = 99,293 \text{ pound-feet}$$

$$\text{Bending Stress Ratio} = \frac{M}{M_n} = \frac{99,293}{134,400} = 0.739$$

$$V = qB(L_c - d) = 2,155 \times 4.00(4.799 - 1.00) = 32,757 \text{ pounds}$$

$$\text{Shear Stress Ratio} = \frac{V}{V_n} = \frac{32,757}{76,800} = 0.427$$

$$q_t = \frac{P+W}{L_{eff}B} = \frac{100,000 + 4,000}{11.599 \times 4.00} = 2,242 \text{ psf}$$

$$\text{Ground Bearing Pressure Ratio} = \frac{q_t}{q_a} = \frac{2,242}{3,000} = 0.747$$

The maximum utilization ratio is found to be 0.747, which is essentially the same as the ratio found by the convergent solution summarized in Table 4.2. The small difference is due to dropping the last term in Eq. 4.18 to produce Eq. 4.19 (the value of this term is small, but not equal to zero). Note that if the solution was exact, then the bending stress ratio and ground bearing pressure ratio would be equal. This small difference is not a concern with respect to practical application of the method.

On the surface, this calculation method appears to be more complex than the older methods outlined in Section 4.1. While this is true, this is not a significant burden. One should recognize that most engineers who perform work of this nature on a regular basis commonly set up the calculations using spreadsheet or mathematics software, such as Microsoft® Excel® or PTC® Mathcad®. Consequently, the extra effort must be expended only once. The reward is a mat design that better represents the behavior of the mat and soil with results expressed as clear utilization ratios.

4.2.2   Performance of the Balanced Mat Analysis Method

The performance of the balanced mat analysis method can be examined by sizing a crane mat using the method and then performing a failure analysis to determine the actual capacity provided. Consider a standard 12" x 4' x 20' hardwood timber crane mat centrally loaded by a 24" wide outrigger float or crawler track. A load of 184,100 pounds (818,920 N) is applied to the mat. The ultimate bearing capacity of the soil is 10,000 psf (480,000 Pa) and a factor of safety of 2.00 is to be applied, giving us an allowable ground bearing pressure of 5,000 psf (240,000 Pa). As before, the allowable stresses for the mat design are 1,400 psi (9,650 kPa) in bending and 200 psi (1,380 kPa) in shear. A summary of the design of the mat based on this effective bearing area method is shown in Table 4.3. As we see, this load brings this mat/soil arrangement to 100% of the mat strength (limited by bending) and ground bearing capacity.

The mat behavior may be treated as linearly elastic for the purpose of this analysis. Soil may be treated as linearly elastic at load levels up to the allowable soil bearing pressure, but it becomes very nonlinear as the ultimate bearing capacity is approached. Thus, we will use two different analysis methods to evaluate this design.

The mat behavior up to the allowable load can be analyzed as a beam on an elastic foundation (Young, et al 2012, Table 8.5, Case 1). In addition to the values defined above, we must also know the modulus of elasticity E of the timbers and the modulus of subgrade reaction k_s of the soil. E may be taken as 1,200,000 psi (8,300,000 kPa) for the wood species commonly used for crane mat construction, as developed in Section 3.5. Bowles (1996) suggests that a practical value of k_s can be calculated using Eqs. 4.37 for the indicated units (these three equations are dimensionally dependent).

$$k_s = 12q_{ult} \qquad (k_s \text{ in kips per cubic foot; } q_{ult} \text{ in ksf)} \quad (4.37a)$$

$$k_s = \frac{q_{ult}}{144} \qquad (k_s \text{ in pounds per cubic inch; } q_{ult} \text{ in psf)} \quad (4.37b)$$

$$k_s = 40q_{ult} \qquad (k_s \text{ in kN per cubic meter; } q_{ult} \text{ in kPa)} \quad (4.37c)$$

TABLE 4.3 Mat Design Results – P = 184,100 lbs. (818,920 N)

	U.S. Cust. Units	SI Units	Comments
Mat Weight, W	4,000 lbs	17,793 N	
L_{eff} (Eq. 4.19)	9.41 feet	2.87 meters	controlling value
L_{eff} (Eq. 4.23)	11.81 feet	3.60 meters	
L_{eff} (Eq. 4.31)	13.54 feet	4.13 meters	
L_c (Eq. 4.33)	3.70 feet	1.13 meters	
q (Eq. 4.32)	4,893 psf	234,278 Pa	
M (Eq. 4.34)	1,610,165 lb-in	181,924 N-m	
f_b (Eq. 4.6)	1,398 psi	9,639 kPa	100% of F_b
V (Eq. 4.35a)	52,903 lbs	235,324 N	
f_v (Eq. 4.8a)	138 psi	951 kPa	69% of F_v
q_t (Eq. 4.36)	5,000 psf	239,401 Pa	100% of q_a

The basis of this approximation for k_s is illustrated in Fig. 4.2. The solid curve is a representation of typical soil load-displacement behavior. The exact shape of this curve is rarely known, so performing "exact" calculations of the deflections of the mat and the soil is not possible. The approximation suggested by Bowles (1996) assumes that the soil deformation is linearly elastic up to the ultimate bearing capacity q_{ult}, beyond which the soil will deform without limit while supporting a constant pressure, as represented by the dashed lines (elastic-perfectly plastic behavior). The change from elastic behavior occurs at a displacement X_{max} assumed to be equal to one inch (25 mm). Thus, k_s is equal to the slope of the line to the left of X_{max} in Fig. 4.2, as expressed in equation form in the figure and in Eqs. 4.37 for specific units.

Thus, given q_{ult} = 10,000 psf = 10 ksf (480 kPa), Eq. 4.37a gives a value of k_s = 12 (10) = 120 kips per cubic foot (69 pci from Eq. 4.37b; in SI units, Eq. 4.37c gives k_s = 40 x 480 = 19,200 kilonewtons per cubic meter).

Given the soil conditions on many construction sites, the use of the approximation for the modulus of subgrade reaction discussed here is questionable. Excavated and backfilled areas, compacted surface layers, and other deviations from a homogeneous soil mass all serve to increase the uncertainty with which soil elastic properties can be determined. The use of this approximation for the purpose of this examination of the design method is reasonable, but its use for actual mat design is generally not recommended for these reasons of uncertainty.

Using these values, the results of the beam on an elastic foundation analysis are shown graphically in Fig. 4.3. We see that the actual bearing pressure between the mat and the soil is greatly variable, not uniform as assumed in the standard design methods. However, the peak bearing pressure due to P is only about 5% greater than the pressure q given by the proposed design method (Table 4.3), which is not significant. The actual bearing length is shown as 15'-1" (4.60 m), markedly greater than the effective bearing length of 9'-5" (2.87 m) calculated using this design method. Other pertinent calculated values are shown in Table 4.4.

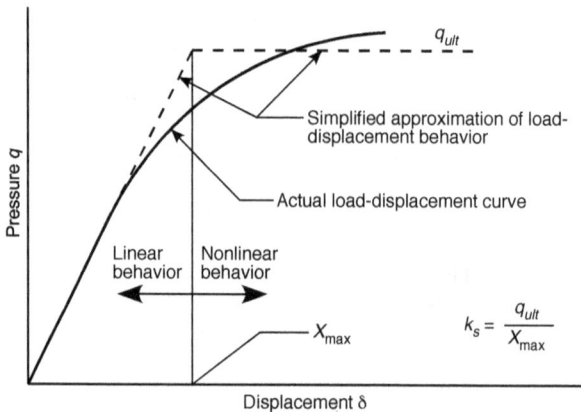

Figure 4.2 Modulus of Subgrade Reaction (based on Bowles 1996, Fig. 9-9)

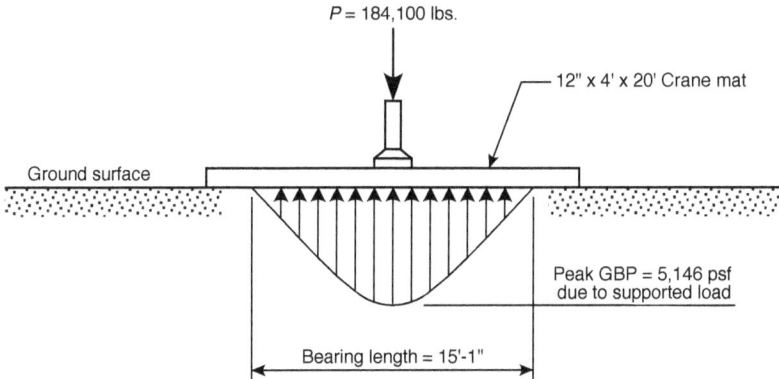

Figure 4.3 Bearing Pressure Curve due to the Supported Crane Load

One value that stands out in Table 4.4 is the bending stress of 1,671 psi, which is 19% greater than the allowable value of 1,400 psi. As we will soon see, this is not significant with respect to the validity of the design method.

We will now proceed with the failure analysis of this mat/soil arrangement. The first analysis is at a load P of 355,375 pounds (1,580,787 N). The ground bearing pressure distribution is shown in Fig. 4.4. The peak ground bearing pressure due to the supported load is 9,934 psf. When the pressure due to the dead weight of the mat is added, the peak ground bearing pressure is found to be 10,000 psf. The bending stress in the mat at this load is 3,226 psi, which is still below the limit state bending stress of 3,400 psi, as discussed in Section 3.5.4.

For the second analysis, we will use the program FADBEMLP, a nonlinear beam-on-soil analysis program that is packaged with the Bowles (1996) text. This program treats the beam as elastic and the soil as elastic-perfectly plastic (the simplified approximation illustrated in Fig. 4.2). The applied load P is taken as 372,000 pounds (1,654,739 N) and the ground bearing pressure distribution is illustrated in Fig. 4.5. We can see from the shape of the pressure curve that the soil still has some additional support capability at this load. The moment in the mat

TABLE 4.4 Mat Elastic Analysis Results – P = 184,100 lbs. (818,920 N)

	U.S. Cust. Units	SI Units	Comments
Mat Weight, W	4,000 lbs	17,793 N	
L_{eff}	15.10 feet	4.60 meters	
L_c (Eq. 4.33)	6.55 feet	2.00 meters	
q	5,146 psf	246,392 Pa	
M	1,925,369 lb-in	217,538 N-m	
f_b (Eq. 4.6)	1,671 psi	11,521 kPa	119% of F_b
V	52,564 lbs	233,816 N	
f_v (Eq. 4.8a)	137 psi	945 kPa	68% of F_v
q_t	5,212 psf	249,552 Pa	104% of q_a

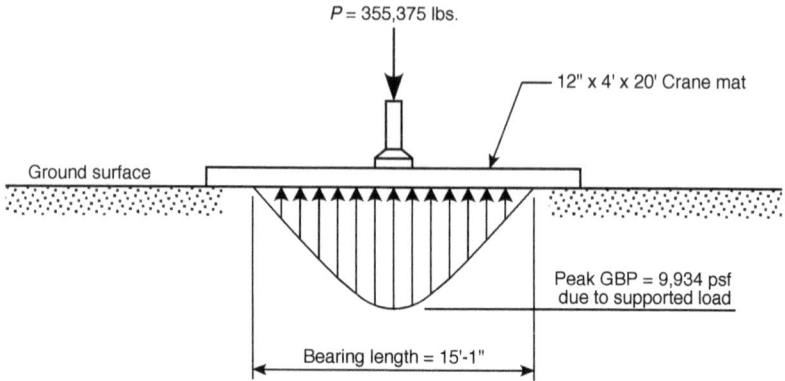

Figure 4.4 Bearing Pressure Curve at 355,375 Pounds (1,580,787 N)

at this load is 3,897,120 pound-inches, which gives a bending stress of 3,383 psi. This indicates that the mat will likely fail before the ultimate bearing capacity of the soil over the full footprint of the mat is reached.

At a soil bearing capacity q_{ult} equal to 10,000 psf (478,800 Pa), the bending strength of the mat governs the ultimate capacity of the mat/soil combination. This can be viewed graphically in Fig. 4.6 as a plot of the applied crane load vs. the mat bending stress, where the bending stress is based on the beam on an elastic foundation analyses in the linear range and on the FADBEMLP analyses in the nonlinear range. Of importance in this figure are the two horizontal lines drawn at the allowable bending stress of $F_b = 1,400$ psi used for mat design based on the common timber species and the limit state bending stress of $F_b = 3,400$ psi, as developed in Section 3.5.4.

We now see that the limit state strength of the crane mat used in these examples is reached at an applied load of 372,000 pounds (1,654,739 N) and that the design method presented here shows an allowable load of 184,100 pounds (818,920 N). Thus, the method provides a design factor of 372,000 / 184,100 = 2.02 on soil with an ultimate bearing capacity of 10,000 psf (478,800 Pa). The limit state of the mat/ soil combination is the bending strength of the mat.

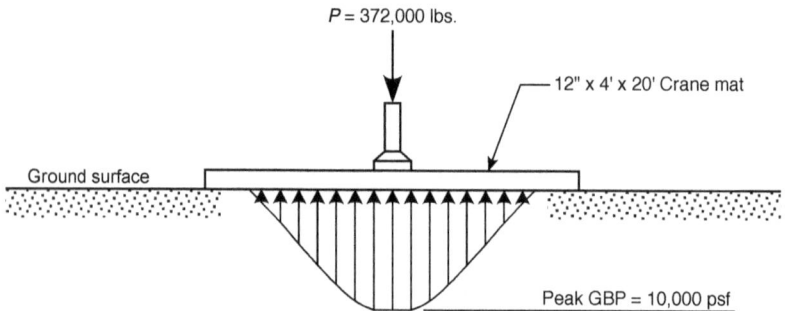

Figure 4.5 Bearing Pressure Curve at 372,000 Pounds (1,654,739 N)

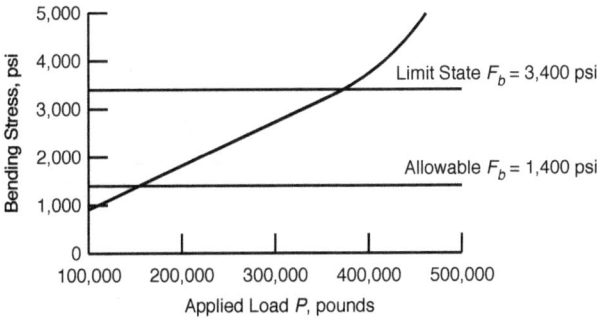

Figure 4.6 Crane Load vs. Mat Bending Stress at $q_{ult} = 10,000$ psf

Soil with an allowable bearing pressure of 5,000 psf (240,000 Pa) is on the high end of the range commonly encountered around much of the U.S. We will now repeat this set of crane mat analyses based on an allowable bearing pressure of 1,500 psf (71,800 Pa). Again using a safety factor of 2.00 for the soil, this value corresponds to an ultimate bearing capacity q_{ult} of 3,000 psf (143,600 Pa).

The same configuration (Fig. 4.1) gives us an allowable crane load P of 90,000 pounds (400,340 N) at the point where the average ground bearing pressure and the mat bending stress both reach their respective allowable values. As we repeat the mat analysis for increasing crane loads, we develop the load vs. bending stress curve shown in Fig. 4.7 (as with Fig. 4.6, the bending stress is based on the beam on an elastic foundation analyses or the FADBEMLP analyses, as applicable).

The peak ground bearing pressure for this mat/soil arrangement reaches q_{ult} at a crane load of 142,500 pounds (633,870 N) and the limit state bending stress is reached at a crane load of 198,000 pounds (880,748 N). The soil is well into the nonlinear range at this load (Fig. 4.8), with the ground bearing pressure equal to q_{ult} along the central 12 feet (3.66 m) of the mat. Thus, the design factor provided for this mat/soil combination is 198,000 / 90,000 = 2.20. Again, the limit state is the bending strength of the mat.

If we check this mat/soil combination using the same soil, but with the allowable ground bearing pressure equal to 1,000 psf (that is, based on a safety

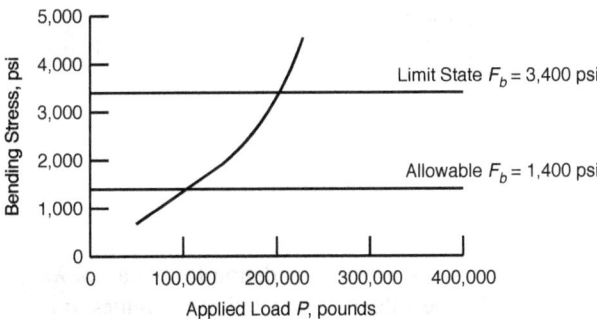

Figure 4.7 Crane Load vs. Mat Bending Stress at $q_{ult} = 3,000$ psf

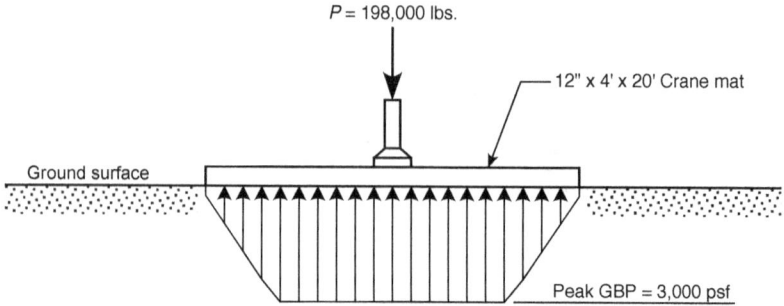

Figure 4.8 Bearing Pressure Curve at 198,000 Pounds (880,748 N)

factor of 3.00), the allowable crane load drops to 71,350 pounds. Since the limit state values remain the same, the design factor becomes 198,000 / 71,350 = 2.78 with this lower allowable ground bearing pressure.

The use of mats that are stronger and stiffer than the standard mat considered in these examples, such as those made with mora timbers, or two layers of mats stacked on top of one another would alter the strength balance such that, in some cases, the ground bearing capacity would become the limit state of the arrangement, rather than the mat strength.

We can see, then, that the true design factor that any of these mat design methods provide is not a constant value, even though the allowable mat stresses and allowable ground bearing pressure are based on fixed design factors. The greater the allowable ground bearing pressure, the lower will be the design factor for a particular size and type of crane mat. However, even at the relatively high value of 5,000 psf (240,000 Pa), the method developed here provides a reasonable design factor of at least 2.02 based on the average bending limit state strength developed in Section 3.5.4 and the design factor is greater lower allowable ground bearing pressures.

4.2.3   Crane Mat Stiffness

A guide related to the relative stiffness of the mat to the soil is given by the value of λL_{eff} as defined in Eq. 4.38 (Bowles 1996). This value is used to distinguish between a rigid and a flexible foundation.

$$\lambda L_{eff} = \sqrt[4]{\frac{k_s B L_{eff}^4}{4 E I_x}} \tag{4.38}$$

The mat is considered to be a rigid foundation for values of λL_{eff} less than $\pi/4$ = 0.79. It is considered to be a flexible foundation for values of λL_{eff} greater than $\pi = 3.14$. Values between these two limits indicate a foundation of intermediate stiffness. The mat example of Table 4.3 gives us a value of λL_{eff} equal to 2.01, thus indicating a stiffness in the intermediate range. Examination of numerous mat designs using q_{ult} from 1,000 psf to 20,000 psf and a depth of 12" (305 mm) generally shows values of λL_{eff} in the range of 1.7 to 2.9 for a single layer of conventional hardwood or softwood timber mats, 1.8 to 3.1 for a single layer of mora mats, and 2.3 to 4.0 for a single layer of high strength (e.g., emtek®) timber mats. The lower values of λL_{eff} occur at the higher values of q_{ult}. Using two layers of mats placed on top of one another with the timbers oriented in the same direction doubles the bending strength and stiffness of the mat assembly. This results in an increase in L_{eff} and a corresponding increase in λL_{eff}, in some cases changing the classification from intermediate to flexible at lower values of q_{ult}. (This assumes that the mats are long enough that L_{eff} remains less than or equal to the actual mat length L_{mat}.)

A comparison of the design factors calculated to support the discussion in Section 4.2.2 shows us that the greater design factors occur in mat/soil combinations that yield greater values of λL_{eff}. This relationship is illustrated in Fig. 4.9 for mat designs based on two different allowable ground bearing pressure safety factors. The curves of Fig. 4.9 are only approximate and are useful primarily as reference values. The mat design examples used to create Fig. 4.9 are based on timber mats in thicknesses of 8", 10" and 12", q_{ult} values of 3,000 psf, 6,000 psf, 8,000 psf and 10,000 psf, and allowable ground bearing pressures based on a safety factor of 2.00 or 3.00. In all cases, the actual mat length is greater than the effective bearing length and the effective bearing length is limited by bending strength.

The relationship between the design factor and λL_{eff} is less well defined for mat designs in which L_{eff} is equal to the actual mat length L. In examples of this configuration that were checked, the design factor of the mat/soil combination was consistently equal to just over 2.00 when the allowable ground bearing pressure was based on a safety factor of 2.00. When mat designs were computed using an allowable ground bearing pressure based on a safety factor of 3.00, the relationship with λL_{eff} was scattered, with the capacity of the combination sometimes based on the strength of the mat and sometimes based on the bearing capacity of the soil. In

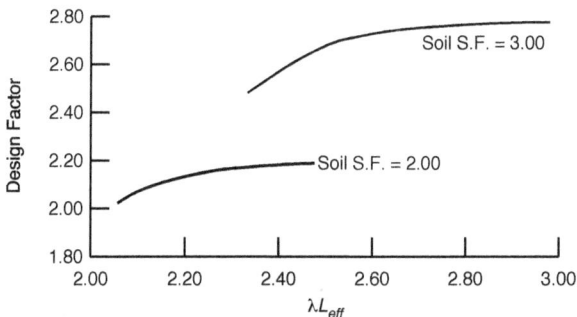

Figure 4.9 Timber Mat Design Factor vs. λL_{eff}

all cases, however, the design factor of the mat/soil combination fell in the range of 2.50 to 3.05. The lower values, of course, were for those combinations in which the strength of the mat was the limiting value.

As with other results that depend on soil elastic properties, this calculation of λL_{eff} should be considered as a guide only when exercising engineering judgment in the solution of a crane mat design problem. The value of λL_{eff} is generally not to be used as a design criterion due to the uncertainty with which the modulus of subgrade reaction k_s is known.

4.2.4   Comments on Support Deflection

The deflection calculation discussed on page 139 provides some insight into the behavior of the mat, but is not an accurate calculation of the true mat/soil deformation under load. This is due to the use of the idealized values of q and L_c in Eq. 4.29. More likely deflections of the mat/soil combination can be investigated using the beam on an elastic foundation approach if a reliable value of k_s is available.

Analyses of dozens of mat/soil combinations with crane loads set equal to the maximum value permitted by the effective length calculation method developed here show that the actual displacement of the mat will differ, sometimes significantly, from that computed using Eq. 4.29. This study examined 12" thick by 4' wide timber mats of oak, mora and high strength laminated wood (emtek®), one or two layers thick, supported on soil with q_{ult} varying from 2,000 psf to 16,000 psf. The soil remained in the elastic range for values of λL_{eff} equal to about 2.25 and below. Above 2.25, a nonlinear analysis is necessary to better evaluate the mat/soil behavior.

The elastic analysis deformations differed from the values calculated using Eq. 4.29. The greatest differences occurred at the lowest values from Eq. 4.29. However, all of the elastic analysis deformations for this group of mat/soil combinations were in the range of about 1/2" (13 mm) to just over 3/4" (19 mm). This indicates that the mat deflections calculated using Eq. 4.29 are not a true indicator of the mat behavior, but that the method developed here will provide reasonable and consistent support to the crane.

We can relate this deflection of the mat/soil combination to the levelness of the crane (which is ultimately what is important) as follows. Consider a crawler track with a triangular bearing pressure envelope (for example, Fig. 1.14) for which the length of the bearing pressure envelope is 20 feet, or 240 inches. Assume that the mat deflection at the heavily loaded end of the track is 0.75", the high end of the deflection range observed in the study discussed above. The deflection at the opposite end where the track bearing pressure of 0 psi is 0". Thus, the slope of the track is 0.75" in 240", which is a slope of 0.31%, or 0.18°. The slope of an outrigger-supported crane can be analyzed in the same manner, with a similarly small slope found.

4.2.5   Eccentrically Loaded Mats

All of the examples shown in this chapter for the development of the effective bearing area calculation method were illustrated with mats that were centrally loaded. This, of course, is not always the case. Crane mats are often loaded eccentrically.

A common situation in which this occurs is shown in Fig. 4.10. Two runs of 20-foot (6.1 m) wide mats are laid side by side under a crawler crane with a track spacing of, say, 24 feet (7.3 m). The result is that each crawler track is offset from the center of the mat. How does this eccentric loading affect the ground bearing pressure imposed by the mat and, consequently, how does it affect the effective bearing length to be used for practical mat analysis?

We can examine this problem by working with the arrangement shown in Fig. 4.10. As shown in the figure, each crawler track is offset from the center of the mat by 2 feet (0.6 m). Let us assume that the load from the crawler track to one mat is 96,000 pounds (427,000 N). We will also assume that the soil has an ultimate bearing capacity of 4,000 psf (192,000 Pa) and we can work with an allowable ground bearing pressure of 2,000 psf (96,000 Pa). Using the approximation of the modulus of subgrade reaction discussed on page 143, we can analyze the actual distribution of the ground bearing pressure using the beam on an elastic foundation model. The result is shown in Fig. 4.11. We see, as expected, that the pressure distribution is asymmetrical. The peak bearing pressure is found to be 2,166 psf (103,709 Pa).

A common method of addressing this mat configuration and loading is to assume that only a length of mat that is symmetrical about the centerline of the crawler track acts to support the crane and that the bearing pressure under this symmetrical length is uniform. That is, the maximum effective bearing length of the mat is $2 L_a$, where L_a is the shorter distance from the centerline of the crawler track (or outrigger float) to the end of the mat, as shown in Fig. 4.10. This approach can be illustrated by means of two examples.

The first example is based on the arrangement shown in Fig. 4.10, in which the centerline of the crane load is 2 feet (0.6 meter) eccentric to the center of the mat, so

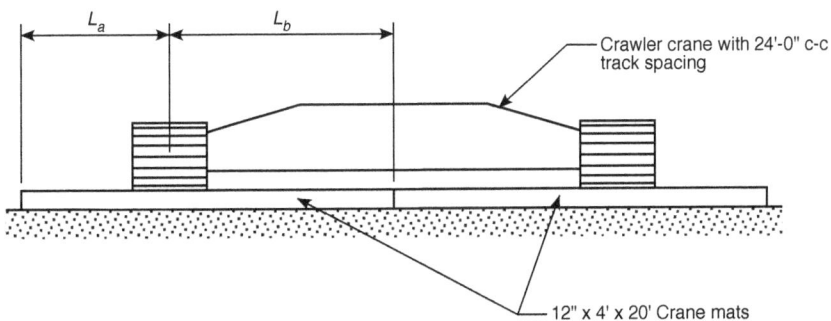

Figure 4.10 Eccentrically Loaded Crane Mats

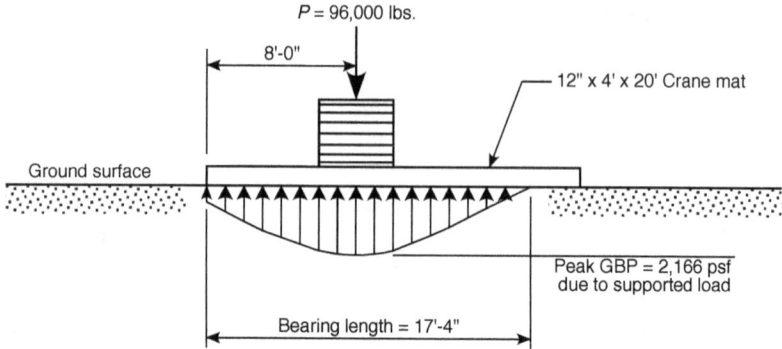

Figure 4.11 Bearing Pressure Distribution Under an Eccentrically Loaded Crane Mat

$L_a = 8$ feet (2.4 meters). As with the previous examples, the width of hard bearing between the crane and the mat is taken as 24" (610 mm). The resulting ground bearing pressure distribution calculated using the beam on an elastic foundation method is shown in Fig. 4.11, as previously noted. The general pressure pattern is as was seen for the symmetrically loaded mat, except the pressure envelope is now somewhat asymmetrical, as would be expected. The results of the mat design calculations based on a symmetric effective bearing length are summarized in Table 4.5. The effective bearing length L_{eff} is 13.80 feet (4.21 m) and the ground bearing pressure due to the crane load q is 1,739 psf (83,264 Pa). The peak ground bearing pressure determined by the elastic analysis is about 25% greater than this average pressure.

The second example will use the same basic arrangement, except the mat length is only 16 feet (4.88 m), thus resulting in an eccentricity of the crane load to the center of the mat of 4 feet (1.22 m) and $L_a = 4$ feet (1.22 meters). The allowable

TABLE 4.5 Analysis of Eccentrically Loaded Mat – Eccentricity = 2.00 feet

	U.S. Cust. Units	SI Units	Comments
Mat Weight, W	4,000 lbs	17,793 N	
L_a	8.00 feet	2.44 meters	
Symmetrical Length	16.00 feet	4.88 meters	
L_{eff} (Eq. 4.19)	13.80 feet	4.21 meters	controlling value
L_{eff} (Eq. 4.23)	16.00 feet	4.88 meters	
L_{eff} (Eq. 4.31)	16.00 feet	4.88 meters	
L_c (Eq. 4.33)	5.90 feet	1.80 meters	
q (Eq. 4.32)	1,739 psf	83,264 Pa	
M (Eq. 4.34)	1,453,305 lb-in	164,202 N-m	
f_b (Eq. 4.6)	1,262 psi	8,701 kPa	90% of F_b
V (Eq. 4.35a)	34,090 lbs	151,640 N	
f_v (Eq. 4.8a)	89 psi	614 kPa	45% of F_v
q_t (Eq. 4.36)	1,811 psf	86,711 Pa	91% of q_a

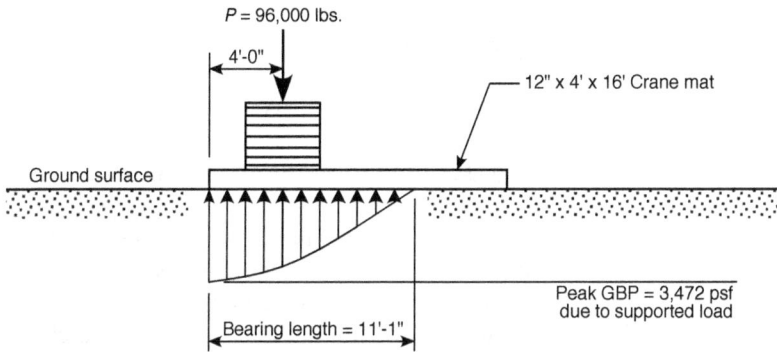

Figure 4.12 Bearing Pressure Distribution Under an Eccentrically Loaded Crane Mat

ground bearing pressure is taken as 3,000 psf (144,000 Pa) for this configuration [q_{ult} = 6,000 psf (288,000 Pa)].

The calculation results for this second example are illustrated in Fig. 4.12 and summarized in Table 4.6. The asymmetry of the pressure pattern is now significant due to the greater offset of the supported crane load. The effective bearing length defaults to the actual symmetrical length for all three limit states. A beam on an elastic foundation analysis of this configuration shows a peak ground bearing pressure of 3,472 psf, about 16% greater than the average pressure q.

Although these examples appear to show an inadequacy in this approximation, this is not the case. As with the symmetrically loaded mats, nonlinear failure analyses of eccentrically loaded crane mats using a range of assumed soil properties show that the soil and the mats have sufficient excess capacity that this simple approximation is a reasonable approach that provides overall design factors that are consistent with the values previously discussed for symmetrically loaded mats.

TABLE 4.6 Analysis of Eccentrically Loaded Mat – Eccentricity = 4.00 feet

	U.S. Cust. Units	SI Units	Comments
Mat Weight, W	3,200 lbs	14,234 N	
L_a	4.00 feet	1.22 meters	
Symmetrical Length	8.00 feet	2.44 meters	controlling value
L_{eff} (Eq. 4.19)	8.00 feet	2.44 meters	
L_{eff} (Eq. 4.23)	8.00 feet	2.44 meters	
L_{eff} (Eq. 4.31)	8.00 feet	2.44 meters	
L_c (Eq. 4.33)	3.00 feet	0.91 meter	
q (Eq. 4.32)	3,000 psf	143,641 Pa	
M (Eq. 4.34)	648,000 lb-in	73,214 N-m	
f_b (Eq. 4.6)	563 psi	3,882 kPa	40% of F_b
V (Eq. 4.35a)	24,000 lbs	106,757 N	
f_v (Eq. 4.8a)	63 psi	434 kPa	32% of F_v
q_t (Eq. 4.36)	3,100 psf	148,429 Pa	103% of q_a

As always, of course, experienced-based professional judgment is occasionally needed to confirm that any particular design approach is suitable for the specific circumstance at hand.

4.2.6   Significance of the Modulus of Subgrade Reaction

The evaluations of the different mat configurations discussed in the preceding sections are all based on values of the modulus of subgrade reaction k_s that have been calculated using the approximation given in Bowles (1996) and discussed starting on page 143. This raises an obvious question: How does a change in the value of k_s affect the results of these beam on an elastic foundation analyses?

This question can be answered quite simply by analyzing crane mats using a range of values of k_s and comparing the results, particularly the bending stress in the mats and the peak ground bearing pressure. A short study has been performed in which a crane mat was sized using the method developed in Section 4.2 and then analyzed as a beam on an elastic foundation using k_s calculated with Eq. 4.37, 0.85 k_s, and 1.15 k_s. This was done for values of the ultimate ground bearing pressure q_{ult} of 3,000 psf, 5,000 psf, 10,000 psf, and 15,000 psf.

In general, this +/- 15% variation in the value of k_s resulted in a variation of the mat bending stress of about +/- 5%. The bending stress variance was slightly greater at higher values of q_{ult}. The peak ground bearing pressure under the center of the mat varied by about +/- 4%, with the variance relatively constant across the range of values of q_{ult}. This indicates that the used of the Bowles (1996) approximation for the value of k_s does not introduce a significant error in the analyses used to evaluate the performance of the design methods. It is noted, of course, that the actual modulus of subgrade reaction of a particular soil mass may differ from the value calculated using Eq. 4.37 by a greater percentage.

4.3   MATS OF OTHER MATERIALS

While crane mats made of rough sawn timbers are by far the most common type of mats used under mobile cranes, they are not the only type. Cranes at both extremes of the rated load spectrum often can benefit from the use of mats made of other materials.

The support loads developed by very high capacity cranes, both crawler and outrigger-supported, often are greater than can be effectively distributed over a large enough area by timber mats. This has driven the use of mats made of structural steel members. At the other end of the spectrum are smaller cranes, particularly rough-terrain cranes, that require mats to obtain adequate support, but can benefit from something more rugged and more easily handled than timber mats. This need has opened the door to the development of mats made of synthetic materials. These products are discussed in this section.

4.3.1   Structural Steel Mats

The outrigger loads of high capacity [over 400 tons (365 tonnes)] cranes may be in excess of 300,000 pounds (1,335,000 N). Given a typical construction site allowable ground bearing pressure on the order of 4,000 psf (191,500 Pa), a mat area of about 75 square feet (7 square meters) will be required under each outrigger float. Building up a mat area of this size with timber mats is fairly impractical. High track bearing pressures under similarly high capacity crawler cranes present the same load spreading challenge. As noted in Chapter 3, this has driven the use of fabricated structural steel crane mats.

The behavior of a steel crane mat is very much the same as the behavior of a timber mat. Therefore, the same process of calculating the effective bearing area is appropriate, using Eqs. 4.16 through 4.36. A review of steel mats designed by the author for both crawler cranes and outrigger-supported cranes finds that the designs almost always fall into the intermediate stiffness range, as defined by Eq. 4.38. The only exception found was one design for an unusually long mat for use under a crawler crane, the stiffness of which fell into the flexible range. For reference, the deflection limit defined by Eqs. 4.29 through 4.31 still was met by this design.

Engineers in the U.S. commonly base allowable stress (or strength) values for crane mat design on the provisions of the AISC *Specification for Structural Steel Buildings* (AISC 1989 or AISC 2016), also as discussed in Chapter 3. Comparable steel design standards used in other countries for building structures must be evaluated by the responsible engineer to determine their suitability for mat design.

In general, then, the structure-on-soil behavior of a fabricated steel crane mat is essentially the same as that of a timber mat. Unlike timber mats, a steel mat can be detailed, as shown in Fig. 4.13, to provide load spreading stiffness in both orthogonal directions (a primary advantage of steel mats), but the fundamental soil-structure interaction is the same.

4.3.2   Synthetic Mats

At the opposite end of the rated load spectrum, small and medium mobile cranes develop outrigger loads that are usually too great to be supported by the crane's floats alone when the crane is set up on bare soil, but an exceptionally large bearing area is generally not required. Mobile cranes of up to 100 tons (90 tonnes) rated load typically develop maximum outrigger loads that are no greater than about 125,000 pounds (556,000 N). The actual outrigger loads that will occur during routine lifting operations, where the loads lifted are comfortably under the crane's rated load, will be markedly less. In these situations, smaller mats are usually adequate.

Crane owners often outfit their cranes with mats that are suitable for use under the most common operating conditions. While wood mats are still common,

Top and bottom plates
welded to all beams with
fillet, plug, or slot welds ———

Fully welded moment
connections at all joints ———

CL of I-shape beams (typ.)

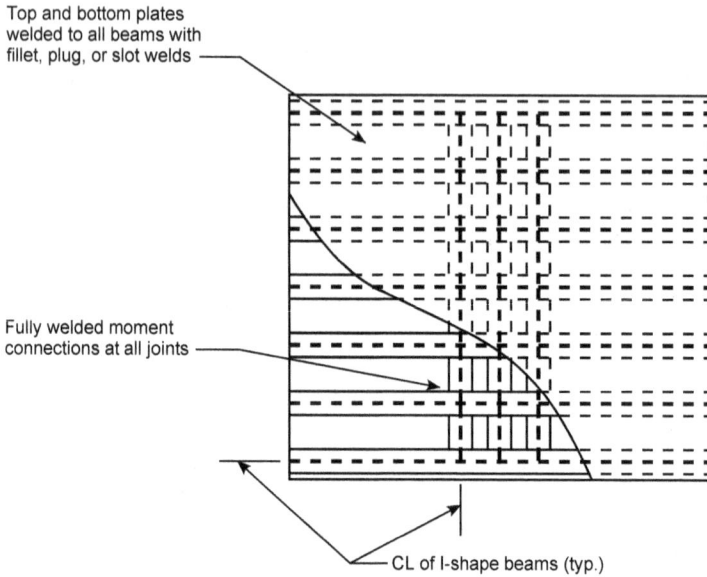

Figure 4.13 Steel Grillage Outrigger Mat

synthetic outrigger pads (Fig. 4.14) are becoming increasingly popular. These pads, which are proprietary products, are most commonly round or square, are available in a variety of sizes (footprint area and thickness) to provide compatibility with expected outrigger loads and ground conditions, and are fitted with grabs or other appurtenances to facilitate easy handling. The pad manufacturer can also specify a maximum allowable outrigger load that the pad can safely support.

Synthetic outrigger pads are typically made of ultra high molecular weight (UHMW) plastic or fiber-reinforced polymer (FRP). The exact material properties often are not published by the pad manufacturers, as they understandably treat their designs as proprietary. Further, a strength analysis of such a pad is generally not a task that a crane user could reasonably perform. When calculating the strength of a timber mat or a fabricated structural steel mat, the load path through the individual members and connections is relatively easy to understand and analyze. This behavior lends itself to analysis using the classical methods with which every structural or mechanical engineer is familiar. The more complex deformed shape that a pad such as those pictured in Fig. 4.14 will take under load is much more difficult to analyze. While a manufacturer can undertake this effort, along with full-scale load testing if needed, as a part of product development, a crane owner or user will find the effort unreasonable.

The bottom line when using synthetic crane mat products like these is one often heard: Abide by the manufacturer's load ratings. For example, if a particular outrigger pad is rated by the manufacturer for an outrigger load no greater than 50,000 pounds (222,400 N), then that is the maximum outrigger load that the

Figure 4.14 Synthetic Outrigger Pads *(DICA® Outrigger Pads)*

pad can safely support. Any additional analysis of the pad by the crane user is unnecessary. Special applications will have to be checked and approved by the manufacturer.

The evaluation of the ground bearing pressure below these synthetic outrigger pads is performed just as it is for timber or steel mats, but using the actual mat area unless otherwise indicated in the product's specifications.

EXAMPLE 4-4

Demonstrate the evaluation of synthetic outrigger pads.

Information from the crane's specifications:
    Maximum crane outrigger load = 38,900 pounds
    Outrigger float bearing area = 346 square inches

Allowable ground bearing pressure = 4,000 psf (from a site geotechnical study)

Outrigger pad specifications from the pad manufacturer:
    Outrigger pad size = 4'-0" x 4'-0" x 1" thick (square)
    Outrigger pad self-weight = 80 pounds
    Outrigger pad rated load = 135,000 pounds
    Outrigger pad allowable bearing pressure (from the outrigger float) = 400 psi

Outrigger pad utilization = 38,900 / 135,000 = 0.29 < 1.00 O.K.
Outrigger float bearing pressure = 38,900 / 346 = 112 psi
Bearing pressure utilization = 112 / 400 = 0.28 < 1.00 O.K.
Total load to supporting surface = 38,900 + 80 = 38,980 pounds
Pad bearing area = 4.0 x 4.0 = 16.0 square feet

Ground bearing pressure = 38,980 / 16.0 = 2,436 psf
Ground bearing pressure utilization = 2,436 / 4,000 = 0.61 < 1.00 O.K.

Conclusion: The outrigger pad and the ground bearing pressure are both acceptable for the stated load and site.

4.3.3   Steel Plates

In some applications, steel plates may be placed directly on the soil for use as crane mats. Plates used in this manner under a crawler crane's tracks may be analyzed in the same manner as is a timber mat using the balanced mat analysis method developed in Section 4.2. The analysis of a plate used under an outrigger float is somewhat more complex. This design problem can be analyzed as a plate on an elastic foundation, if the software to perform such an analysis is available. This analysis will result in a peak bearing pressure under the center of the plate, as was observed in the results of the beam on an elastic foundation problems examined earlier in this chapter.

Recognizing that many engineers do not have access to software that can perform this plate on an elastic foundation analysis, the recurring problem of determining an accurate value of the modulus of subgrade reaction, and that common practice in mobile crane installation engineering calls for evaluation based on a uniform bearing pressure, a different approach is desirable.

The problem of an outrigger float supported on a steel plate bearing directly on soil can be solved with reasonable accuracy using readily available handbook formulas. Solutions will vary somewhat, depending on the float and plate details and on the handbooks used. However, the concept of this type of solution can be demonstrated by means of an example based on the support shown in Fig. 4.15.

EXAMPLE 4-5
Analyze the steel plate mat shown in Fig. 4.15 using formulas from Young et al (2012) and Hsu (1990).

Outrigger load = 50,000 pounds
Outrigger float dimensions - 24" x 24"
Steel plate dimensions - 1 1/2" x 60" x 60"
Steel grade - ASTM A36
F_y = 36 ksi
E = 29,000 ksi
Allowable ground bearing pressure = 2,500 psf

Neither reference addresses this problem directly, so the solution is developed by combining two cases. These are a plate simply supported on four sides and loaded by a uniform pressure over its entire area (Young et al 2012 Table 11.4, Case 1a) and a plate

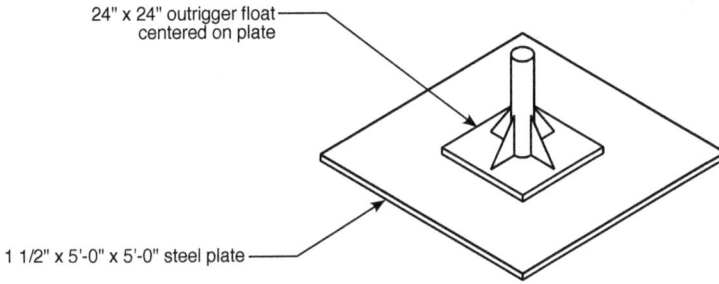

Figure 4.15 Steel Plate Under an Outrigger

simply supported on four sides and loaded by a uniform pressure over a defined centrally positioned rectangle (Young et al 2012 Table 11.4, Case 1c, and Hsu 1990 Case 3-5). The superposition of these two cases approximates the behavior of a plate loaded by a uniform pressure over its entire area on one side and loaded by a uniform pressure over a defined centrally positioned rectangle on the other side (Fig. 4.16).

Applying Young et al (2012) Table 11.4, Case 1a, we can solve for both the bending stress and deflection. The notation used here is based on that shown in Young et al (2012) Table 11.4.

$a/b = 60/60 = 1.0$
$\beta = 0.2874$
$\alpha = 0.0444$

$$q = \frac{50,000}{60(60)} = 13.89 \text{ psi} = 0.0139 \text{ ksi}$$

$$f_b = \frac{\beta q b^2}{t^2} = \frac{0.2874(0.0139)60^2}{1.50^2} = -6.39 \text{ ksi (tension on the top)}$$

$$\Delta = \frac{\alpha q b^4}{Et^3} = \frac{-0.0444(0.0139)60^4}{29000(1.50^3)} = -0.082 \text{ inch}$$

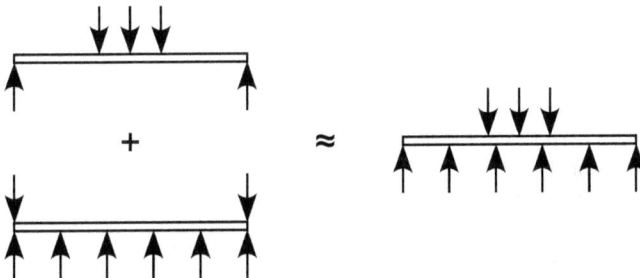

Figure 4.16 Superposition of Two Plate Analysis Cases

Applying Young et al (2012) Table 11.4, Case 1c, we can solve for the bending stress. This table doesn't address deflection for this case, so we must apply Hsu (1990) Case 3-5 to calculate the deflection. The notation in these two handbooks is not the same, so care must be used to assure that the correct values are used in each equation.

$a/b = 60/60 = 1.0$

$a_1/b = 24/60 = 0.4$

$b_1/b = 24/60 = 0.4$

$\beta = 0.84$

$$f_b = \frac{\beta W}{t^2} = \frac{0.84(50.0)}{1.50^2} = 18.67 \text{ ksi (tension on the bottom)}$$

Combined bending stress $f_b = -6.39 + 18.67 = 12.28$ ksi
Allowable bending stress $F_b = 0.75\,F_y = 27.00$ ksi
Bending Stress Ratio = 12.28 / 27.00 = 0.45 O.K.

Now using the Hsu (1990) notation to solve for deflection...

$b/a = 60/60 = 1.0$

$u/a = 24/60 = 0.4$

$v/a = 24/60 = 0.4$

$K_w = 10.26$

$$\Delta = \frac{Pa^2}{100Et^3}K_w = \frac{50.0(60^2)}{100(29000)1.50^3}10.26 = 0.189 \text{ inch}$$

Combined deflection $\Delta = -0.082 + 0.189 = 0.107$ inch
Diagonal length across corners of plate = 84.85 inches
Diagonal length across corners of outrigger float = 33.94 inches
Maximum cantilevered length of plate = (84.85 - 33.94) / 2 = 25.46 inches
Allowable deflection = 0.75% of cantilevered length
　　　　　　　= 0.0075 x 25.46 = 0.191 inch
Deflection Ratio = 0.107 / 0.191 = 0.56 O.K.

Through-thickness shear stress, although rarely (if ever) a limiting performance criterion, should be checked. The shear area A_v is the perimeter of the outrigger float multiplied by the plate thickness.

$A_v = 1.50 (24 + 24 + 24 + 24) = 144$ in.^{2}

$f_v = (50,000 / 1,000) / 144 = 0.35$ ksi

Allowable shear stress $F_v = 0.40\,F_y = 14.40$ ksi

Shear Stress Ratio = 0.35 / 14.40 = 0.02 O.K.

Weight of steel plate = 1,531 pounds
Ground bearing pressure = (50,000 + 1,531) / (5.0 x 5.0) = 2,061 psf

Ground Bearing Pressure Ratio = 2,061 / 2,500 = 0.82 O.K.

Bending stress, shear stress, deflection, and ground bearing pressure are all acceptable. Thus, the plate is shown to be acceptable for this outrigger load, float size, and support condition. If one or more checks showed the plate to be inadequate, the calculations can be repeated using reduced plate dimensions to arrive at an effective bearing area.

4.4   ALTERNATE MAT ARRANGEMENTS

All of the material developed in this chapter to this point is based on the use of a single layer of mats under the crane's outriggers or crawler tracks. This is the basic and most common crane mat configuration, but by no means the only configuration. Mats may be layered on top of one another, either with all of the timbers running in the same direction or with the timbers perpendicular to each other from one layer to the next, steel plates may be laid over timber mats, etc. This section presents discussions of the behavior of the five most common alternate mat arrangements.

The common thread in these five subsections is the development of an understanding of how the loads imposed by the crane are distributed among the mats, plates, and ultimately to the ground surface. Once that understanding is achieved, the principles discussed here for these specific arrangements can easily be modified and applied to other support arrangements. The fundamental requirement imposed upon the lift planner/engineer is to truly understand the principles at work in mobile crane support and not just crunch the numbers through the various equations.

4.4.1   Multiple Layers of Mats – Mats Parallel

A single layer of mats occasionally is not adequate. This is particularly true when a large crane is set up on poor soil. A common solution is to build up a crane pad using two or more layers of mats. The mats may be arranged with all timbers parallel to one another, or with the timbers of one layer perpendicular to those of the adjacent layer. The first case is considered in this section and the second case is addressed in Sections 4.4.2 and 4.4.3.

A common condition for the support of a crawler crane is shown in Fig. 4.17. Two or more layers of mats are laid on top of one another, with the timbers of all layers parallel to one another and perpendicular to the length of the crawler track. The fundamental question to be answered when evaluating the load spreading provided by this arrangement is: How is the load imposed by the crane shared among the layers of mats? (And to answer a common misconception first, two 6" thick mats laid on top of one another does not provide the bending strength of a single 12" thick mat.)

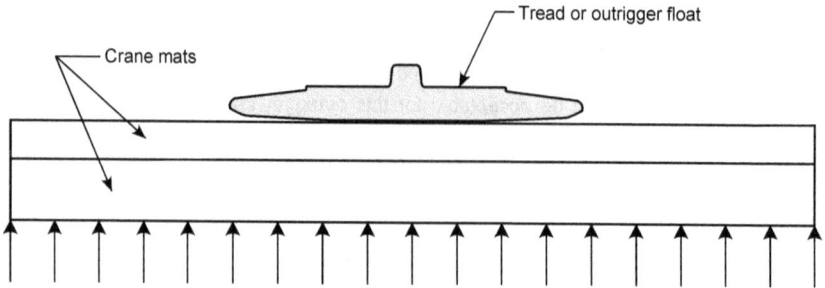

Figure 4.17 Two Layers of Mats – Timbers Parallel

The answer to this question is simple. The load imposed by the crane is shared among the layers of mats in proportion to the bending stiffness of each mat layer, where the bending stiffness of each mat layer is equal to its moment of inertia multiplied by its modulus of elasticity. This load distribution is shown in equation form as Eq. 4.39.

$$L_x = L_t \frac{E_x I_x}{\sum_{i=1}^{i=n} E_i I_i} \qquad (4.39)$$

where

L_x	=	crane load to mat layer x;
L_t	=	total crane load;
E_x	=	modulus of elasticity of mat layer x;
I_x	=	moment of inertia of mat layer x; and,
n	=	total number of mat layers.

Eq. 4.39 is solved for each mat layer. The total of each value of L_x so calculated must add up to L_t.

The simplest version of this mat arrangement is two layers of identical mats. In this case, EI is the same for both layers and Eq. 4.39 is solved to show that each layer receives a load $L_x = 0.50 L_t$. Thus, the effective bearing length equations in Section 4.2.1 can be solved using strength values (allowable shear and allowable moment) equal to twice the strength of one mat and a moment of inertial equal to twice that of one mat. And fortunately, not only is this the simplest version of this mat arrangement, it is also the most common.

If two layers of crane mats of different thicknesses are used, as is illustrated in Fig. 4.17, the following procedure is used to determine the effective length.

1. Calculate the fraction L_x of the total crane load L_t that will be carried by each mat layer using Eq. 4.39.
2. Let the maximum value of $L_x = L_{max}$ for mat layer x.
3. Calculate $pct = L_{max} / L_t$.

4. Calculate $M_n = M_{nx} / pct$, $V_n = V_{nx} / pct$, and $(EI)_n = \sum (E_i I_i)$.
5. Use these values of M_n, V_n, and $(EI)_n$ in the equations of Section 4.2.1 to determine the effective bearing length of the mats.
6. Once the effective bearing length has been determined, check the resulting moment, shear, and deflection in both layers. Note that the deflections of both of the mat layers must be equal to one another (thus indicating that the mats remain in contact with one another as they flex under load).

Application of this design procedure for a two-layer arrangement of 8" mats over 12" mats is illustrated in Example 5-3.

The analysis becomes somewhat more complex if the upper mats are shorter than the lower mats. In this case, both the strength and the stiffness are not constant over the full cantilevered length L_c of the mats. The solution process can follow the same procedure as outlined above, but the checking of the moment and shear must be done at the point of maximum loading for each different number of layers. Likewise, the deflection must be calculated taking into account the differing number of layers.

In most practical configurations, this process will give the correct effective bearing length. However, if the strength or deflection checks show an overloading of one or more mat layers, the effective bearing length must be shortened manually and the strength and deflection checks repeated. In this situation, the problem is solved iteratively. Again, this highlights the value of setting up the equations in a math or spreadsheet program, thus permitting rapid and accurate iteration.

The illustration of this mat arrangement in Fig. 4.17 implies the use of two or more layers of timber crane mats. However, since Eq. 4.39 is written in terms of both the modulus of elasticity and the moment of inertia of each mat layer, the procedure outlined here is not limited to timber mats. By using appropriate values of E and I, the behavior of, for example, a steel mat over timber mats can be solved for the load distribution where the steel mat load spreading is in the same direction as the timbers.

4.4.2   Multiple Layers of Mats – Mats Perpendicular – Case 1

The second mat arrangement based on two or more layers that we will address is that in which the timbers of one layer are set perpendicular to the timbers of the adjacent layer. This mat arrangement can be used with both crawler cranes and outrigger-supported cranes, although the calculation method differs somewhat, depending on the type of crane.

We will first examine the use of this configuration under a crawler crane. A common mat layout is illustrated in Fig. 4.18. The lower layer of mats is laid out with the timbers perpendicular to the length of the crawler track, as is normally done. The upper layer of mats is positioned with the timbers oriented parallel to the crawler track length. The primary function of these upper (longitudinal)

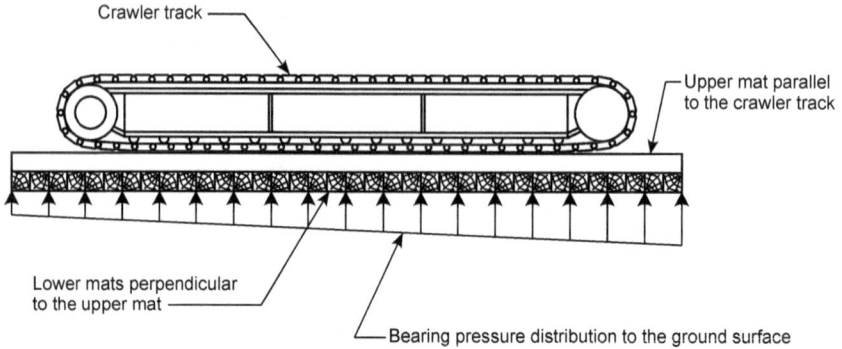

Crawler track

Upper mat parallel to the crawler track

Lower mats perpendicular to the upper mat

Bearing pressure distribution to the ground surface

Figure 4.18 Two Layers of Mats – Timbers Perpendicular – Case 1

mats is to increase the effective bearing length of the crawler tracks to the lower mats. In summary, the track bearing pressure is applied to the top surface of the longitudinal mat and a corresponding bearing pressure envelope on the underside of the mat is calculated. The most important requirement in this calculation is that the centroid of the track bearing pressure envelope must be coaxial with the centroid of the underside bearing pressure envelope (Fig. 4.19).

The calculation of the underside bearing pressure envelope is performed just as one would calculate the ground bearing pressure under a spread footing that is loaded by an eccentrically located vertical load. This problem is commonly treated in foundation engineering textbooks (e.g. Bowles 1996). When applied to crane mats, however, the solution is only valid when the cantilevered length is short and/or stiff enough that deflection of the mat is small.

The first step is to resolve the track bearing pressure envelope into a vertical load P and the location of P relative to the more heavily loaded end of the track c'. This is done using Eqs. 4.40 and 4.41. The position of P can also be expressed as an eccentricity e_t, as defined by Eq. 4.42. The notation used in these equations is illustrated in Fig. 4.20.

$$P = \frac{\left(t_{max} + t_{min}\right)}{2} CL \tag{4.40}$$

$$c' = \frac{3Lt_{min} + L\left(t_{max} - t_{min}\right)}{3\left(t_{max} + t_{min}\right)} \tag{4.41}$$

$$e_t = \frac{L}{2} - c' \tag{4.42}$$

Note that if the minimum track bearing pressure t_{min} is zero (that is, if the track pressure envelope is triangular), Eq. 4.41 reduces to $c' = L / 3$ and Eq. 4.42 reduces to $e_t = L / 6$. Although not used in the following calculations, the moment due to the eccentricity is simply $P (e_t)$.

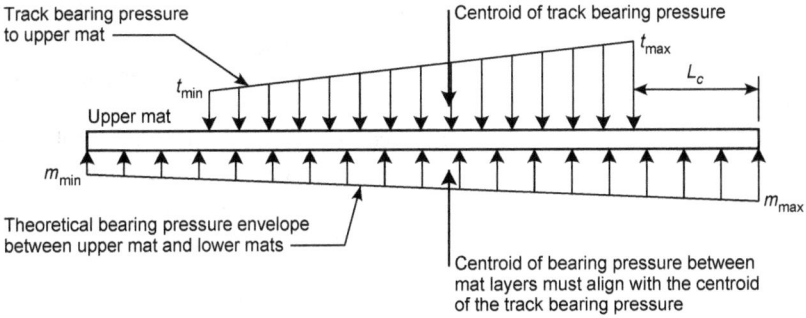

Figure 4.19 Two Layers of Mats – Timbers Perpendicular – Upper Mat Loading

With P and c' calculated, Eqs. 4.43 through 4.46 are used to calculate the pressure envelope acting on the underside of the upper mat. Note that m_{max} and m_{min} are given in units of force per unit length. If the shape of the mat bearing pressure envelope is triangular, Eq. 4.46 correctly calculates m_{min} as zero.

$$L_{eff} = 3\left(c' + L_c\right) \le L_{mat} \tag{4.43}$$

$$e_m = \frac{L_{eff}}{2} - \left(c' + L_c\right) \tag{4.44}$$

$$m_{max} = \frac{P}{L_{eff}}\left(1 + \frac{6e_m}{L_{eff}}\right) \tag{4.45}$$

$$m_{min} = \frac{P}{L_{eff}}\left(1 - \frac{6e_m}{L_{eff}}\right) \tag{4.46}$$

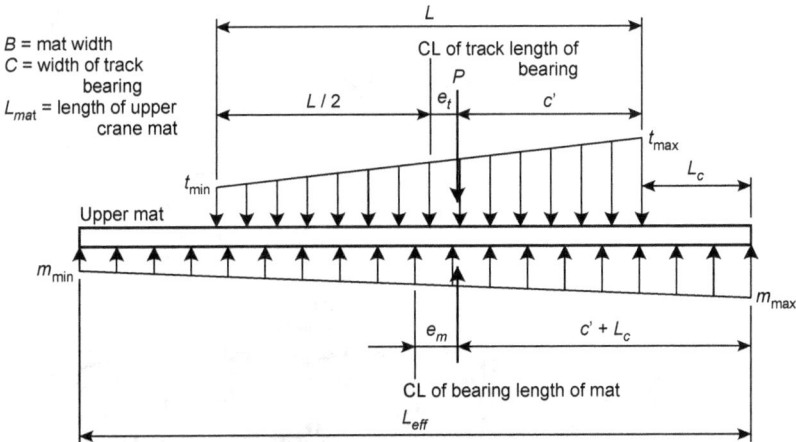

Figure 4.20 Upper Mat Loading Notation

The values calculated with Eqs. 4.40 through 4.46 are then used to calculate the shear, moment, and deflection in the cantilevered length of the upper mat. If the resulting stresses and deflection are acceptable, the use of L_c based on the geometry illustrated in Figs. 4.19 and 4.20 is acceptable. If not, the calculations must be repeated using a value of L_c that is reduced as required to give acceptable stress and deflection results. An allowable deflection equal to 0.75% of L_c is appropriate. Note that since we don't have a clearly defined allowable bearing pressure that can be applied to the top of the lower mats, we can't calculate an effective L_c using a closed solution, as is shown in Section 4.2.1 for a single layer of mats.

The performance of an upper mat layer in this arrangement is best illustrated by means of an example. Fig. 4.21 is an output sheet from Liebherr's LICCON program for a particular lift made with a Liebherr LR1400/2 crawler crane. The right side track bearing pressures shown are the worst case for this lift. The track bearing length is 7,800 mm (25.6 feet) and the track bearing width is 1,040 mm (40.94 inches). The track is centered on the length of a 30-foot (9.1 m) long mat. The analysis is detailed in Example 4-6. The results are illustrated in Fig. 4.22.

Although not shown in the example, the pressure due to the dead weight of the upper mat must be added to the pressures due to the crane loading.

EXAMPLE 4-6

Calculate the loads acting on the top and bottom surfaces of the upper mats.

$L = 7,800$ mm $= 307.09"$
$C = 1,040$ mm $= 40.94"$
$t_{max} = 41$ psi x $40.94" = 1,679$ pounds per inch

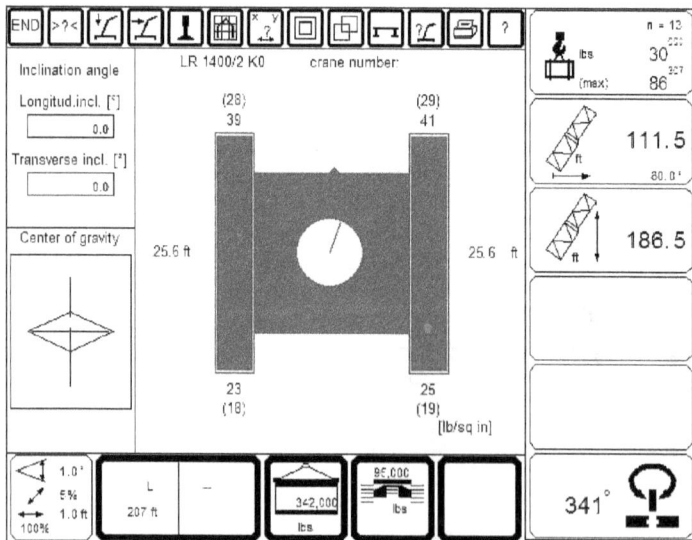

Figure 4.21 Liebherr LR1400/2 Track Bearing Pressures

t_{min} = 25 psi x 40.94" = 1,024 pounds per inch

From Eq. 4.40, P = 414,930 pounds
From Eq. 4.41, c' = 141.14"
From Eq. 4.42, e_t = 12.41"
From Eq. 4.43, L_{eff} = 360.00" (= the full length of the mat L_{mat})
$L_c = (L_{eff} - L) / 2 = (360.00 - 307.09) / 2 = 26.46"$
From Eq. 4.44, e_m = 12.41"
From Eq. 4.45, m_{max} = 1,391 pounds per inch
From Eq. 4.46, m_{min} = 914 pounds per inch

Some of the numerical results summarized above and illustrated in Fig. 4.22 may appear to be off. This is due to independent rounding.

With these load values now determined, the moment is calculated at L_c and the shear at $L_c - d$ for a timber mat or at L_c for a steel mat. The deflection of the more heavily loaded end should also be calculated. In a case like this, where L_c is only slightly more that two times the mat thickness, the deflection will be very small. Provided that the mat stresses and deflection are acceptable, this pressure pattern is the loading applied to the lower layer of mats. The lower layer of mats is analyzed using the previously established method.

One final point must be made about the practical functionality of this mat arrangement. The use of an upper layer of mats placed parallel to the length of the crawler tracks to increase the crawler track bearing length generally only works if the crane is to remain stationary. If the crane must travel, the upper mats lose their effectiveness as the more heavily loaded end of the crawler track approaches the end of the upper mat. When the heavy end of the track reaches the end of the mat, the load spreading effect of the upper mat is mostly or completely lost. This effect is illustrated in Fig. 4.23. Here we see the track bearing pressure envelope acting on the top surface of the mat and the pressure under the mat in this position. When the track pressure envelope is triangular (Fig. 4.23a), the pressure under the

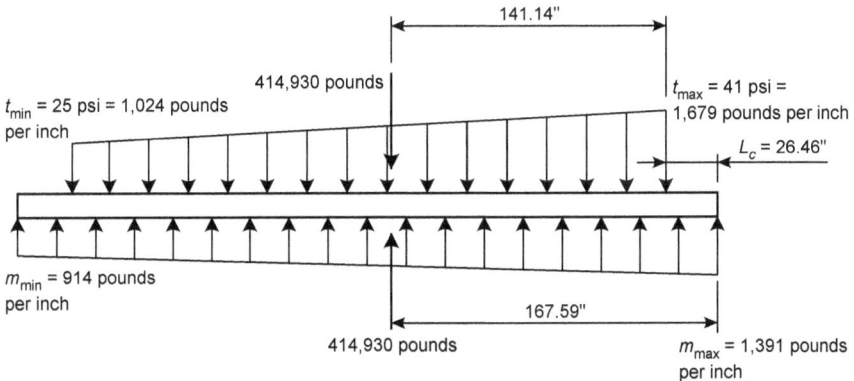

Figure 4.22 Distribution of LR1400/2 Track Bearing Pressure by Upper Mat

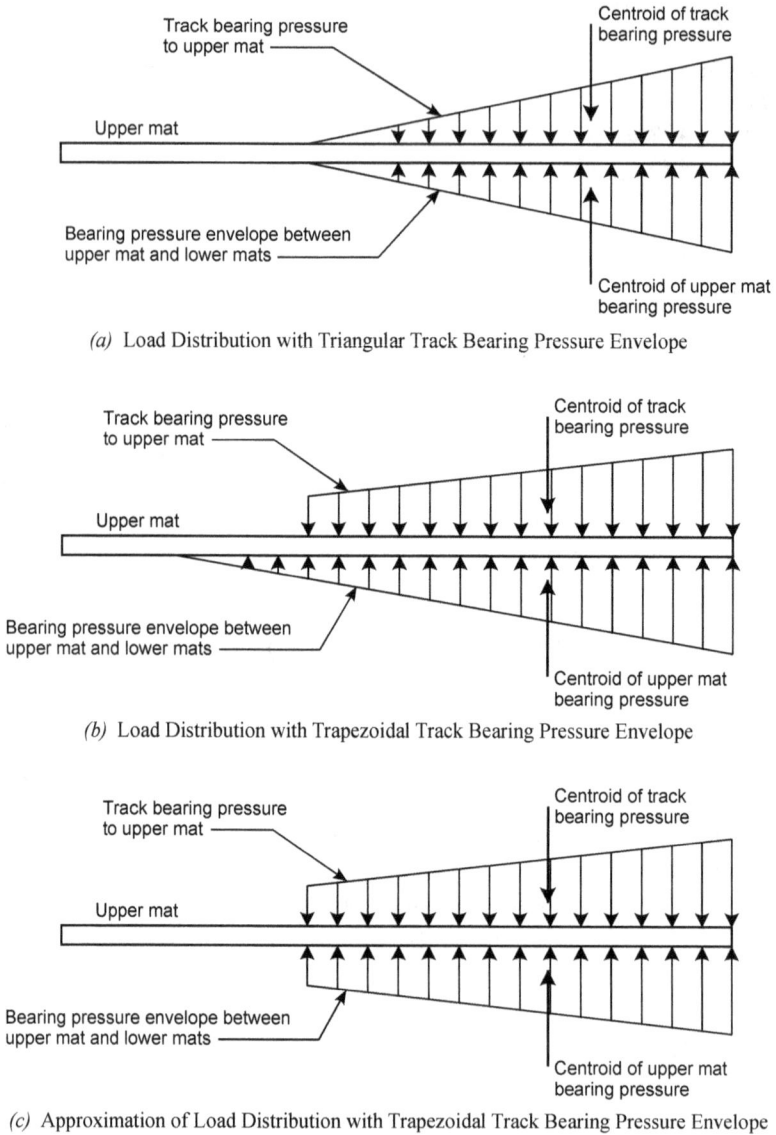

(a) Load Distribution with Triangular Track Bearing Pressure Envelope

(b) Load Distribution with Trapezoidal Track Bearing Pressure Envelope

(c) Approximation of Load Distribution with Trapezoidal Track Bearing Pressure Envelope

Figure 4.23 Upper Mat Load Distribution with Crawler at the End of the Mat

mat will be identical in form, so there is no longitudinal load spreading. When the
track bearing pressure envelope is trapezoidal (Fig. 4.23*b*), the maximum bearing
pressure on the underside of the mat will actually be somewhat greater than the
track bearing pressure above if the mat is long enough to allow development of the
indicated triangular pressure envelope below (remember that the centroids of the
upper and lower pressure envelopes must be coaxial). For practical application,

however, it is often reasonable to assume an absence of change in the pressure envelope (Fig. 4.23c).

A fundamental characteristic of the track and mat pressure envelopes was made on page 163 and in the preceding paragraph. This point is very important and is worth repeating. The centroid of the track bearing pressure envelope on the top surface of the longitudinal mat and the centroid of the bearing pressure envelope of the mat to the surface below must be coaxial. If they do not align, there is an error in the calculations. Verifying this alignment provides a practical check of the bearing pressure calculations.

Two layers of timber mats set perpendicular to each other may also be used under an outrigger to obtain load spreading in both directions. This analysis is fairly straightforward, employing the equations already established. In summary, the upper mat is analyzed using the usual assumption of a uniform bearing pressure on the bottom surface. The effective bearing length is established using the considerations of bending strength, shear strength, and deflection discussed in the preceding sections. The load and footprint from this mat is then used as the loading applied to the lower mats and, again, they are analyzed using the method derived in Section 4.2.1.

4.4.3   Multiple Layers of Mats – Mats Perpendicular – Case 2

Another two-layer mat arrangement in which the mat timbers are perpendicular to each other is illustrated in Fig. 4.24. Here, the upper layer of mats are arranged with the timbers perpendicular to the length of the crawler tracks and the mats in the lower layer are arranged with the timbers parallel to the length of the tracks. As with the mat arrangement discussed in Section 4.4.2, the goal again is to obtain load spreading in two directions.

The analysis of the upper layer of mats is performed exactly as discussed in Section 4.2.1 for a single layer of mats on soil. The only question in the application of these calculations is what value to use for the allowable bearing pressure q_a. As

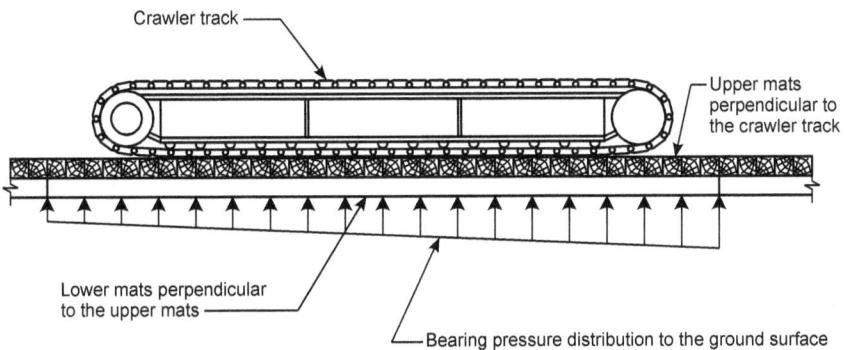

Figure 4.24 Two Layers of Mats – Timbers Perpendicular – Case 2

we will see, this is another strong argument for performing these calculations using spreadsheet or math software.

One product of the analysis of the upper mat layer is a footprint (length and width) and bearing pressure envelope. These values define the loading to the lower mat layer. These mats are analyzed in a manner similar to that used on the upper mat layer in the Case 1 arrangement discussed in Section 4.4.2. And as with the Case 1 arrangement, if the crane is to travel, the lower mat layer will lose its load spreading effectiveness when the end of the crawler track reaches the end of the mats. In this case, the only load spreading is provided by the upper layer and the analysis of the mats is the same as for single-direction mats on soil.

If we are to obtain a balanced mat design, the maximum stress ratios for both mat layers and the utilization ratio for the ground bearing pressure must be the same. As with the Case 1 mat arrangement, developing equations that provide a closed solution to this design is not practical due to the number of variables involved. A practical method is to set up the equations for both mat layers and then manually adjust the effective bearing lengths until the results converge.

The discussions in Sections 4.4.2 and 4.4.3 are based on only two layers of mats, which is the most common situation. However, we can easily see the use of three or more layers, either with each layer perpendicular to the adjacent layer(s) or various combinations of two or more layers parallel to one another and other layers above or below perpendicular. The same basic mat performance principles apply, but the calculations become significantly more complex due to the need to balance that many more components in the support system. In any of these arrangements, the calculation of the deflections of the mat layers can be a very powerful tool for verifying the accuracy of the strength analysis. The mats will always remain in contact with one another, which means that the deflection of one layer must be equal to the deflection of the next layer. If the calculated deflections are not equal, then something is wrong in the analysis. It is incumbent upon the lift planner/engineer to correctly apply the calculations to arrive at a valid determination of the load spreading provided by the proposed mat arrangement.

4.4.4   Mats Continuous Under Both Crawler Tracks

Timber crane mats are commonly available in lengths up to about 30 feet (9 m) and steel mats can be fabricated to any length, limited primarily by transportation and handling considerations. Thus, a crawler crane with a track gauge of up to about 24 feet (7.3 m) can be set up on a run of these longer mats such that each mat supports both tracks (Fig. 4.25). The analysis of this arrangement is somewhat more complex than that of the mat with a single load previously addressed.

As one would expect, the ground bearing pressure under the mat is greatest directly under each crawler track and then drops off toward the center. The results of an elastic analysis of such a mat is illustrated in Fig. 4.26a. As with the other types of mat-on-soil analyses discussed in the preceding sections, performance of

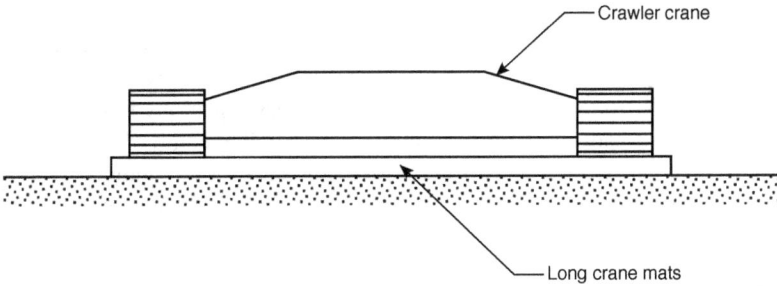

Figure 4.25 Mat Continuous Under Two Crawler Tracks

this type of analysis again requires a reliable value of the soil modulus of subgrade reaction k_s, which typically is not available. Further as previously shown, an elastic analysis doesn't represent the true performance of the mat/soil. Therefore, we need a practical solution that is not dependent on quantities that are not in hand.

A common and simple method of mat analysis is based on the assumption that a segment of the mat between the crawlers is ineffective and that the two lengths of mat under the crawlers impose a uniform bearing pressure on the supporting surface. This load/support model is illustrated in Fig. 4.26*b* and is also briefly presented in Shapiro and Shapiro (2011).

In practice, the length of the ineffective segment of mat in the center is adjusted in the calculations until the greatest stress ratio in the mat (usually bending stress) is equal to the ground bearing pressure utilization ratio. That is, just like the basic mat design approach derived in Section 4.2.1, the design seeks to balance the utilization of mat strength with that of ground bearing capacity. The analysis model of the mat is that of a beam with two short uniformly distributed loads acting on the top surface (the track loads) and two uniformly distributed loads acting on the bottom surface (the ground bearing pressure).

The analysis of a continuous mat presents a number of variables. The two crawler loads may not be equal, the overhangs on either side may not be equal, and

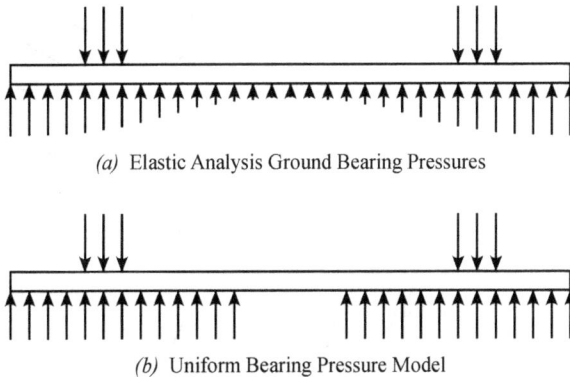

(a) Elastic Analysis Ground Bearing Pressures

(b) Uniform Bearing Pressure Model

Figure 4.26 Continuous Mat Bearing Pressure Distribution

the ineffective segment in the center will vary in length. Given these considerations, this analysis does not lend itself to a straightforward solution, as does the mat with a single load. As with other more complex mat arrangements, the author's approach to solution of this problem is to set up the necessary equations in a spreadsheet and then manually adjust certain variables, primarily the length of the ineffective segment between the tracks, until the results converge. This method has proven itself over time to be a practical means of analyzing a continuous crane mat.

4.4.5   Steel Plate or Synthetic Pad over Timber Mats

A steel plate or a synthetic outrigger pad may be used over a timber mat, as shown in Fig. 4.27. The most common reason for employing a plate or a pad in this manner is to assure that the load from an outrigger with a relatively small float is distributed to all of the timbers of the lower mat. For example, a 22" diameter outrigger float centered on a mat made of four 12" x 12" timbers will bear only on the two inner timbers. As discussed in Section 3.4, the tie rods typically used in crane mat construction may not be strong enough to distribute the load to the two outer timbers.

When a steel plate is used for transverse load distribution in this manner, the analysis of the plate is performed just like the design of the mat, itself. That is, the plate is assumed to be loaded by a uniform pressure and the distance from the edge of the outrigger float to the edge of the mat is the cantilever length L_c. The strength of the plate in bending and shear is calculated using a suitable steel design standard. Engineers in the U.S. typically use AISC (1989) or AISC (2016). Comparable structural steel design standards can be used in other countries. The plate is checked for bending, shear, and deflection. Once the plate has been verified

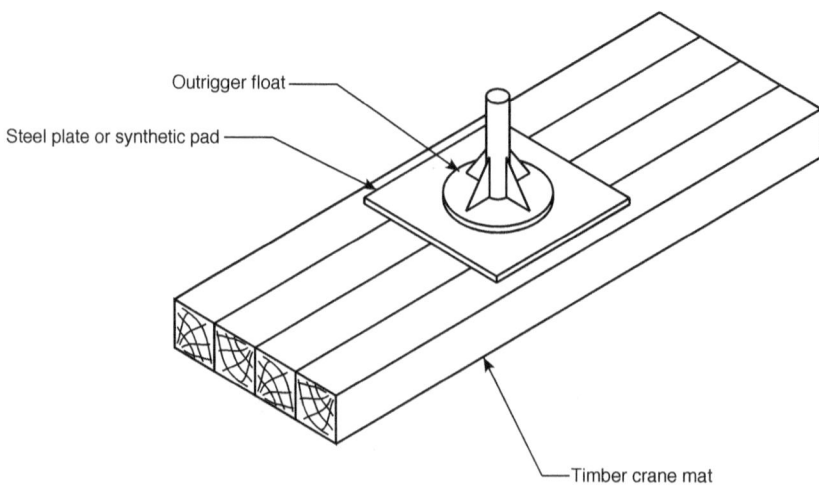

Figure 4.27 Steel Plate or Synthetic Pad Over Timber Mat

as capable of distributing the outrigger load to the timbers, the plate dimensions are used in place of the float dimensions for the timber mat analysis. The timber mat is then checked using the procedure developed in Section 4.2.1. The steel plate design procedure is illustrated in Example 4-7.

EXAMPLE 4-7

Evaluate the transverse distribution of an outrigger load of 75,000 pounds provided by a steel plate laid over a timber mat for the arrangement shown in Fig. 4.28. Evaluate the plate using the allowable stresses given in AISC (1989).

Outrigger float width = 22"
Plate thickness = 1.00"
Plate length = 42.00" (direction across the width of the mat)
Plate width = 40.00" (direction along the length of the mat)
Steel grade - ASTM A36 (F_y = 36 ksi; E = 29,000 ksi)
Mat length = 10'
Mat width = 4'

The ground bearing pressure q due to the outrigger load and the cantilever length L_c to be used in the plate strength checks are as follows.

$$q = \frac{75,000}{4 \times 10} = 1,875 \text{ psf} = 13.02 \text{ psi}$$

$$L_c = \frac{48.00}{2} - \frac{22.00}{2} = 13.00"$$

These values are now used to check the strength and stiffness of the plate.

$$M = \frac{qL_c^2}{2} = \frac{(13.02 \times 120.00)13.00^2}{2} = 132,031 \text{ pound-inches}$$

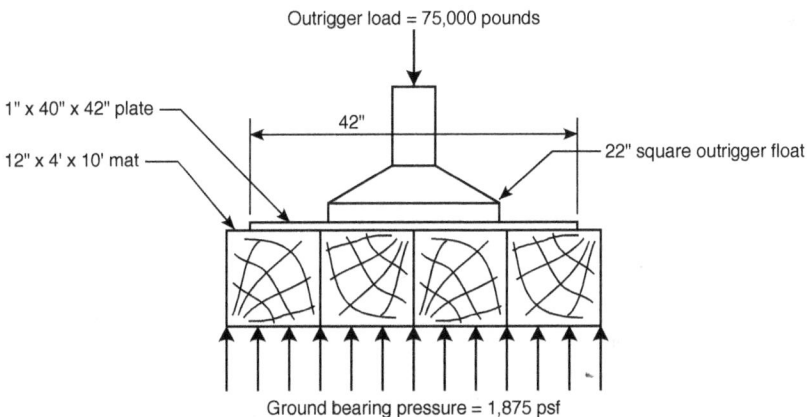

Figure 4.28 Steel Plate on Timber Mat of Example 4-7

$$S_y = \frac{1.00^2 40.00}{6} = 6.67 \text{ in.}^3$$

$$f_b = \frac{132{,}031}{6.67} = 19{,}805 \text{ psi}$$

$$F_b = 0.75F_y = 0.75 \times 36{,}000 = 27{,}000 \text{ psi}$$

$$\text{Bending S.R.} = \frac{f_b}{F_b} = \frac{19{,}805}{27{,}000} = 0.73$$

$$V = qL_c = (13.02 \times 120.00)13.00 = 20{,}313 \text{ pounds}$$

$$f_v = \frac{20{,}313}{1.00 \times 40.00} = 508 \text{ psi}$$

$$F_v = 0.40F_y = 0.40 \times 36{,}000 = 14{,}400 \text{ psi}$$

$$\text{Shear S.R.} = \frac{f_v}{F_v} = \frac{508}{14{,}400} = 0.04$$

$$I_y = \frac{1.00^3 40.00}{12} = 3.33 \text{ in.}^4$$

$$\Delta = \frac{wL_c^4}{8EI} = \frac{(13.02 \times 120.00)13.00^4}{8(29{,}000{,}000)3.33} = 0.058" = 0.44\% \text{ of } L_c$$

The plate is shown to be strong enough and stiff enough to effectively distribute the applied outrigger load to all four timbers of the mat.

A few comments about Example 4-7 are needed.

- First is the question of how to handle a round outrigger float. Calculating L_c from the edge of the float is unconservative. A practical approximation uses a float width equal to 0.707 times the float diameter. If the float in Example 4-7 was round with a diameter of 22", L_c would be taken as 48.00 / 2 - 0.707 x 22.00 / 2 = 16.22". The author has been using this approximation successfully for many years.
- Example 4-7 uses the full width of the plate of 40" as effective in resisting bending. Due to flexure of the mat and plate along the length of the mat, and therefore the width of the plate, the full area of the plate will remain in contact with the mat. Thus, use of the full width of the plate to calculate the cross-sectional properties of the plate is reasonable.
- As indicated in the previous point and illustrated in Example 4-8 starting on page 175, the plate will also be subjected to bending stress along its width. Should these two stresses be combined? A suitable equation for

combining normal stresses like this is the Energy of Distortion formula. However, since both stresses are of the same sense (that is, both stresses are compression on the top of the plate and tension on the bottom of the plate), the Energy of Distortion formula gives a combined stress that is actually less than the greater of the two normal stresses. Therefore, evaluating each bending stress independently results in a conservative solution.

A synthetic outrigger pad used in this function need only be checked by comparing its rated load to the applied outrigger load, as discussed in Section 4.3.2 and demonstrated in Example 4-4. Provided that the pad manufacturer's load rating is not exceeded, no further analysis is necessary by the user to demonstrate acceptability of the pad. The pad dimensions are then used in the analysis of the timber mats below.

Some crane users try to improve the load spreading ability of a timber mat by laying steel plate on top of the mat. In this case, the sharing of the load between the steel plate and the timber mat can be determined by application of Eq. 4.39. This often is not very effective due to the differences in bending stiffness between a plate and a mat, as is illustrated in Example 4-8.

EXAMPLE 4-8

Evaluate the sharing of a crane load between a 12" (305 mm) thick timber crane mat and 1" (25 mm) and 2" (50 mm) steel plates using Eq. 4.39. Both the mat and the plates are 48" (1,220 mm) wide.

E = 1,200,000 psi for timber
E = 29,000,000 psi for steel
I = 6,912 in.^{4} for the timber mat
I = 4 in.^{4} for the 1" steel plate
I = 32 in.^{4} for the 2" steel plate

Eq. 4.39 solved for 1" plate using a total load L_t equal to 100 units:

$$L_x = L_t \frac{E_x I_x}{\sum_{i=1}^{i=n} E_i I_i} = 100 \frac{29,000,000 \times 4}{(29,000,000 \times 4)+(1,200,000 \times 6,912)} = 1.38$$

We see that the 1" plate will carry only 1.38% of the total load.

Eq. 4.39 solved for 2" plate using a total load L_t equal to 100 units:

$$L_x = L_t \frac{E_x I_x}{\sum_{i=1}^{i=n} E_i I_i} = 100 \frac{29,000,000 \times 32}{(29,000,000 \times 32)+(1,200,000 \times 6,912)} = 10.06$$

Here, the steel plate will carry 10.06% of the load. In either case, the resulting bending stress in the plate must be calculated and compared to an appropriate allowable stress.

We can see from this example that laying a steel plate on the top of a timber mat does not present a significant gain in strength and load spreading ability unless the plate thickness is 20% to 25% of the timber thickness or greater. Thinner plates simply do not have the bending stiffness relative to the bending stiffness of the timber mats needed to be effective in providing a significant improvement in load spreading.

One useful application of steel plates over timber mats is for providing load spreading under rubber tires. Fig. 4.29 shows a path of steel plates over timber mats used under a Manitowoc Max-Er® counterweight carrier. Key to the success of this use is assuring through the layout of the mats and plates that adequate support is provided as the tires roll from one plate to the next.

4.4.6   Laminated Wood Mats

Laminated wood mats are widely used on construction sites, particularly in the early stages of a greenfield project, to provide temporary roads for trucks and other rubber-tired equipment. The most common type of laminated mat is that made of three layers of 2" (50 mm) thick boards, as illustrated in Fig. 4.30. Laminated mats in the U.S. are commonly made of the same oak and other common hardwood species that are used for the manufacture of heavy timber crane mats. Because of their ready availability, laminated mats are occasionally suggested for use as crane mats.

The analysis of a laminated mat used as a crane mat combines the methods discussed in the preceding sections for multiple layers of mats, with one layer perpendicular to the others. When used under a crawler crane with the long dimension of the mat perpendicular to the length of the crawler track, the middle layer of boards provides little or no contribution to load spreading. Only the two

Figure 4.29 Steel Plates Under a Max-Er® Wheeled Carrier *(Senco Construction, Inc.)*

layers of longitudinal boards will act and since they are the same thickness, they will share the imposed load equally. Used under an outrigger, all three layers will act to provide load spreading in both directions.

Laminated mats are generally not useful as crane mats (and the laminated mat vendors do not present them as such). The reason why is simple: The 2" thick boards are simply not strong enough or stiff enough in bending to provide significant load spreading. The calculation tools developed in this chapter can be used to fully evaluate a laminated mat, but past experience shows that they are rarely adequate for mobile crane support.

4.5   APPLICATION SUMMARY

This chapter presents the derivations of methods and equations that can be used to analyze the behavior of crane mats in a manner that is practical for routine use for lift planning and engineering. The most commonly encountered crane support configurations are addressed, but there are obviously many other situations that will arise in practice that are not directly covered here. For this reason, it is necessary for the reader to develop a solid understanding of the principles that underlie the methods. In this way, these principles can be used to expand the methods developed here to cover any other mobile crane support problem that may be encountered.

The material in this chapter may be divided into two groups. These are material that is directly applicable for use in the design or analysis of mobile crane support

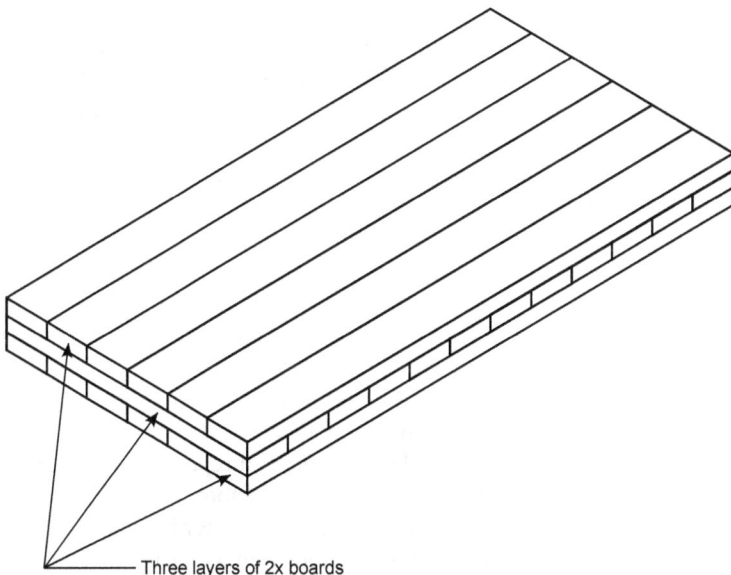

Three layers of 2x boards

Figure 4.30 Laminated Wood Mat

and material that is used to develop or explain the methods and equations that fall into the first group. The reader must understand the difference to assure that only the appropriate material is used in practice.

The most basic mat design scenario is that of a mat (or a group of mats laid side by side) loaded by the crane, either through an outrigger float or a crawler track. Such a mat is analyzed by calculating its effective bearing length using the equations in Section 4.2 that form the Balanced Mat Analysis Method. The equations to be used for mat design when the crane load is centered on the length of the mat are developed in Section 4.2.1. When the crane mats are loaded by the crane placed eccentrically on the length of the mat, the practical solution recommended in this chapter calls for treating the effective bearing length as being symmetrical about the centerline of the crane outrigger or track. This analysis approach is explained in Section 4.2.5.

Sections 4.2.2 through 4.2.4 provide analyses of various aspects of the Balanced Mat Analysis Method that are useful for understanding the approach developed in Section 4.2.1 and to show why this method is a valid means of designing mats for mobile crane support.

As previously stated, the older mat design methods outlined in Section 4.1 are presented for reference and development of the Balanced Mat Analysis Method only and are not recommended for use.

Most of the material in this chapter is based on conventional timber crane mats, simply because timber mats are at present the most common type of mats used in practice by the construction industry. Crane mats also may be fabricated from a variety of other materials, including structural steel and various synthetics. Mats made of these other materials are discussed in Section 4.3 The fundamental behavior of mat bending and mat/soil interaction holds true for all elastic mat materials, so the concepts developed for timber mats apply to these other materials, as well. In the case of a mat made of a proprietary material for which material and section properties are not readily available, direction from the mat manufacturer must be followed.

Section 4.4 addresses a number of relatively common mat arrangements that differ from the single crane load on the single layer of mats covered in Section 4.2. These arrangements include two crane loads on a continuous mat, multiple layers of mats with all timbers parallel to one another, and multiple layers of mats with the timbers of the top layer perpendicular to the timbers of the lower layer(s). Appropriate equations are presented to provide methods of analysis of these arrangements. These equations are applicable to mats made of any structural material, not just timber. The analysis of more complex mat arrangements can be developed following the basic principles applied here.

As has been shown in the example problems and repeated many times in the text, the true behavior of the mat/soil combination is more complex than is implied by the calculation approaches developed in this chapter. While a more theoretically "exact" approach that accounts for this interaction is easily performed using software tools that are widely available, such an analysis usually is not practical due to the difficulty in determining reliable values for the elastic properties of the soil, particularly the modulus of subgrade reaction. As a result, it is sometimes necessary to apply engineering judgment in the solution of a mobile crane support design problem. Because of this potential need, it is necessary that users of this material possess the engineering background and practical experience required to exercise this judgment.

4.6 REFERENCES

Aluminum Association (AA) (2015), *Specification for Aluminum Structures*, Arlington, VA.

American Institute of Steel Construction (AISC) (1989), *Specification for Structural Steel Buildings – Allowable Stress Design and Plastic Design*, Chicago, IL.

American Institute of Steel Construction (AISC) (2016), *Specification for Structural Steel Buildings*, Chicago, IL.

American Wood Council (AWC) (2016), ANSI/AWC NDS-2015 *National Design Specification® for Wood Construction* (NDS), Leesburg, VA.

Bowles, J.E. (1996), *Foundation Analysis and Design*, 5th ed., The McGraw-Hill Companies, Inc., New York, NY.

Duerr, D. (2010), "Effective Bearing Length of Crane Mats," presented at the Crane & Rigging Conference, Houston, TX, Maximum Capacity Media, LLC, Fort Dodge, IA.

Hsu, T.H. (1990), *Structural Engineering & Applied Mechanics Data Handbook, Volume 3: Plates*, Gulf Publishing Company, Houston, TX.

Shapiro, L.K., and Shapiro, J.P. (2011), *Cranes and Derricks*, 4th ed., The McGraw-Hill Companies, Inc., New York, NY.

Young, W.C., Budynas, R.G., and Sadegh, A.M. (2012), *Roark's Formulas for Stress and Strain*, 8th ed., The McGraw-Hill Companies, Inc. New York, NY.

5 Mobile Crane Support Design

The singular reason this handbook was written can be summed up in one sentence: The loads imposed by a mobile crane on its supporting surface often exceed the local strength of that surface. We know that this gap between loading and capacity is bridged by crane mats. This brings us to two questions. What, exactly, is demanded of the support of a mobile crane? And, how does the lift planner/engineer go about fulfilling those demands?

The design of a crane mat requires a determination of the loads from the crane that is to be supported (discussed in Chapter 1), an assessment of the support capacity of the underlying surface, usually soil but possibly some type of structure (discussed in Chapter 2), an assessment of the strength and stiffness of the mats, whether timber or another material (discussed in Chapter 3), and a calculation of the area of the mat that actually bears on the surface and contributes to the support of the crane (discussed in Chapter 4).

This final chapter pulls together the methods and principles developed in Chapters 1 through 4 and summarizes the process needed to design the crane support. Again, it is emphasized that this discussion should not be viewed as a crane support design "cookbook." Rather, this is a description of a design process that can be followed as presented for many crane support applications or modified when necessary for other applications to produce a safe and practical mobile crane support design. The reader must go back through the detailed material in the first four chapters when necessary to modify the calculations in order to adapt the methods to situations not explicitly addressed here.

5.1 REGULATORY LANGUAGE

The need to provide a mobile crane with a firm and level support is required in the U.S. by regulations and construction industry safety standards applicable to mobile crane use. The following paragraphs, quoted from 29 CFR 1926.1402, outline the OSHA requirements for the support of a mobile crane (OSHA 2018).

§ 1926.1402 Ground conditions.
(a) *Definitions.*
(1) "Ground conditions" means the ability of the ground to support the equipment (including slope, compaction, and firmness).

(2) "Supporting materials" means blocking, mats, cribbing, marsh buggies (in marshes/wetlands), or similar supporting materials or devices.

(b) The equipment must not be assembled or used unless ground conditions are firm, drained, and graded to a sufficient extent so that, in conjunction (if necessary) with the use of supporting materials, the equipment manufacturer's specifications for adequate support and degree of level of the equipment are met. The requirement for the ground to be drained does not apply to marshes/wetlands.

(c) The controlling entity must:

(1) Ensure that ground preparations necessary to meet the requirements in paragraph (b) of this section are provided.

(2) Inform the user of the equipment and the operator of the location of hazards beneath the equipment set-up area (such as voids, tanks, utilities) if those hazards are identified in documents (such as site drawings, as-built drawings, and soil analyses) that are in the possession of the controlling entity (whether at the site or off-site) or the hazards are otherwise known to that controlling entity.

(d) If there is no controlling entity for the project, the requirement in paragraph (c)(1) of this section must be met by the employer that has authority at the site to make or arrange for ground preparations needed to meet paragraph (b) of this section.

(e) If the A/D [assembly/disassembly] director or the operator determines that ground conditions do not meet the requirements in paragraph (b) of this section, that person's employer must have a discussion with the controlling entity regarding the ground preparations that are needed so that, with the use of suitable supporting materials/devices (if necessary), the requirements in paragraph (b) of this section can be met.

(f) This section does not apply to cranes designed for use on railroad tracks when used on railroad tracks that are part of the general railroad system of transportation that is regulated pursuant to the Federal Railroad Administration under 49 CFR part 213 and that comply with applicable Federal Railroad Administration requirements.

The following definition from 29 CFR 1926.1401 is also significant in the context of this discussion.

Blocking (also referred to as "cribbing") is wood or other material used to support equipment or a component and distribute loads to the ground. It is typically used to support lattice boom sections during assembly/disassembly and under outrigger and stabilizer floats.

With respect to blocking use under outrigger and stabilizer floats, crane mats fall under this definition of blocking.

The following paragraphs are quoted from ASME B30.5-2018 (ASME 2018).

5-3.2.1.5 Moving the Load

(i) When outrigger floats are used, they shall be attached to the outriggers. Blocking under outrigger floats, when required, shall meet the following requirements:

(1) sufficient strength to prevent crushing, bending, or shear failure

(2) such thickness, width, and length as to completely support the float, transmit the load to the supporting surface, and prevent shifting, toppling, or excessive settlement under load

(3) use of blocking only under the outer bearing surface of the extended outrigger beam

5-3.4.7 Footing

Firm footing under both crawler tracks, all tires, or individual outrigger pads should be level within 1%. Where such a footing is not otherwise supplied, it should be provided by timbers, cribbing, or other structural members to distribute the load so as not to exceed the allowable bearing capacity of the underlying material.

ASME B30.5-2018 also defines certain responsibilities with respect to the support of the crane. The Site Supervisor (the party that has supervisory control over the work site) is responsible for ensuring that the area for the crane is adequately prepared. This includes consideration of surface conditions, levelness, support capability, underground utilities, and the like. The Lift Director (the person who directly oversees the work being performed with the crane) is responsible for ensuring that preparation of the work area has been completed. Last, the Crane Operator must know the types of site conditions that could adversely affect the crane operation and consult with the Lift Director concerning the possible presence of those conditions.

The operator's manual of each crane also defines support requirements that must be met prior to operation of the crane. Anywhere that the crane manufacturer's support requirements are more stringent than those defined in the standards and regulations, the manufacturer's requirements govern.

Note that the provisions given in the OSHA regulations and in ASME B30.5 define only general performance requirements for crane support. The engineering methods by which these requirements are to be met remain the responsibility of those using the crane (the Crane Operator or the Lift Director, depending on the complexity of the crane support requirements).

5.2   REQUIRED INFORMATION

The design of mobile crane supports requires two sets of input information. These are the loads imposed by the crane, whether through outriggers or crawler tracks, and the allowable bearing pressure of the supporting surface. The first of these two items, the crane support loads, is often calculated or otherwise determined by the same individual who is designing the crane installation. The second item, the surface support capacity, is often provided by others. Both of these sets of input information are addressed in this section.

The methods and principles discussed in Chapters 1 through 4 will be demonstrated by means of an example mobile crane support design problem, presented as Examples 5-1 through 5-4 in this chapter. The initial description of the lift to be made will be outlined in the following section and then the design will be built upon in subsequent sections.

5.2.1   Crane Support Loads

Calculation of the crane loads to be carried by the supports must be done with reasonable accuracy.   Many crane manufacturers now provide tools, either stand-alone applications or spreadsheet templates, that the lift planner can use to compute the support loads for their products for a lift configuration that is defined by the user. These tools should be used wherever possible. These applications are discussed in Section 1.3.

The example design problem developed in this chapter begins here. Example 5-1 illustrates the calculation of the track bearing pressures for the specified crane and lift.

EXAMPLE 5-1

Calculate the crawler track bearing pressures imposed by the crane on its supporting surface for the following lift:

Crane – Manitowoc 16000 Series 3 crawler crane
Boom length – 177 feet (54 meters)
Load – 100,000 pounds (45,360 kilograms)
Radius – 75 feet (23 meters)
Possible swing – 360 degrees
Operating surface for bearing pressure calculation – Hard (timber crane mats)

Calculate the track bearing pressures using the Manitowoc Ground Bearing Pressure Estimator (GBPE) program.   Excerpts of the program output showing the track bearing pressures appear in Fig. 5.1a for the crane making the specified lift, in Fig. 5.1b for the crane boomed up to its minimum radius with no load on the hook, and in Fig. 5.1c for the crane raising the boom from horizontal. The maximum track bearing pressures are found to occur when the crane swings with no load on the hook (Fig. 5.1b).

In this case, the results of the crane load calculations are track bearing pressure envelopes defined by maximum and minimum pressures plus a length of bearing for each track.   When designing supports for a truck, all-terrain or rough-terrain

Project: Track Bearing Pressure Example	Operating Surface: Hard - Tread Contact Width 50 Inches
Model: 16000 Series 3	Primary: 177FT (54M) #58HL BOOM
Machine Counterweight: 332,000-LB (150,590-KG) + 120,000-LB (54,430-KG)	Secondary: 0FT (0M) No Attachment
Crawler:Fixed, Position: Gantry Up	Load: From Boom 100,000 LBS @ 75 ft radius

Imperial Pressure Conversion Factor: 1 PSI = 144 PSF

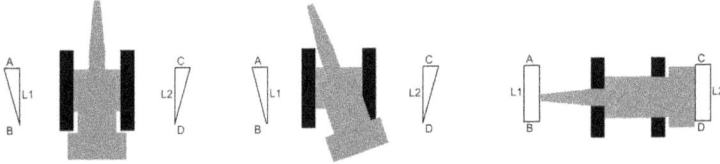

Boom Over Front or Rear	**Boom at Critical 19 degree Swing**	**Boom Over Side**

Center of Rotation to Fulcrum: 148.23 inch		Center of Rotation to Fulcrum: 148.23 inch	Center of Machine to Ctr of Crawlers: 144 inch
A	64.6 psi	66.6 psi	43.0 psi
B	0.0 psi	4.3 psi	43.0 psi
L1	291.7 inch	296.5 inch	296.5 inch
C	64.6 psi	59.2 psi	20.5 psi
D	0.0 psi	0.0 psi	20.5 psi
L2	291.7 inch	281.7 inch	296.5 inch

(a) Track Bearing Pressures for the Specified Lift

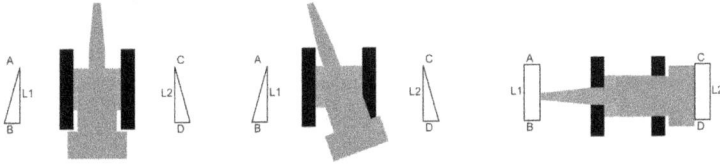

Boom Over Front or Rear	**Boom at Critical 16 degree Swing**	**Boom Over Side**

Center of Rotation to Fulcrum: 148.23 inch		Center of Rotation to Fulcrum: 148.23 inch	Center of Machine to Ctr of Crawlers: 144 inch
A	0.0 psi	0.0 psi	13.3 psi
B	78.9 psi	70.5 psi	13.3 psi
L1	214.4 inch	204.6 inch	296.5 inch
C	0.0 psi	0.0 psi	43.7 psi
D	78.9 psi	81.8 psi	43.7 psi
L2	214.4 inch	237.3 inch	296.5 inch

(b) Track Bearing Pressures with No Hook Load at Minimum Radius

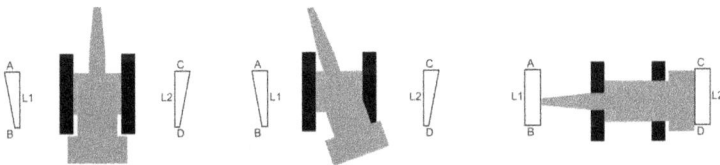

Boom Over Front or Rear	**Boom at Critical 19 degree Swing**	**Boom Over Side**

Center of Rotation to Fulcrum: 148.23 inch		Center of Rotation to Fulcrum: 148.23 inch	Center of Machine to Ctr of Crawlers: 144 inch
A	43.0 psi	43.9 psi	33.5 psi
B	14.0 psi	16.4 psi	33.5 psi
L1	296.5 inch	296.5 inch	296.5 inch
C	43.0 psi	40.6 psi	23.6 psi
D	14.0 psi	13.2 psi	23.6 psi
L2	296.5 inch	296.5 inch	296.5 inch

(c) Track Bearing Pressures While Raising the Boom

Figure 5.1 Manitowoc GBPE Output for Critical Track Pressures

crane, the calculation results will be sets of outrigger loads. The maximum outrigger load is usually the only value of interest, with all outriggers supported in the same manner.

5.2.2 Supporting Surface Strength

The design of a mobile crane installation cannot be made reliably without knowledge of the load carrying ability of the supporting surface, whether that surface is soil, pavement, or some type of structure. Considerations that go into the determination of the allowable ground bearing pressure for a site, including the evaluation of buried pipes and utilities under the effect of surface loading due to a crane, are discussed in Chapter 2. Although this example problem has the crane supported on soil, Chapter 2 also briefly discusses cranes supported on structures.

The allowable soil bearing pressure for the site must be determined by a qualified engineer, typically a specialist in geotechnical engineering. The allowable bearing pressures used for the design of foundations for permanent structures are often based on a factor of safety of at least 3.00. This factor of safety is made necessary by the uncertainties in loading that exist due to the long-term service that the foundation must provide and considerations of consolidation (gradual deformation over time) of the soil. Design of the support of a mobile crane is often based on loads that are relatively well defined and long-term settlement is not an issue for crane set-up. Thus, a lower factor of safety for mobile crane installation design may be appropriate. In the author's experience, an allowable soil bearing pressure based on a factor of safety of 2.00 is often acceptable.

The allowable ground bearing pressure q_a used in the crane support design must consider the strength of subsurface pipes and structures, as well as the strength and stiffness of the soil. Methods of analyzing buried pipes are presented in Chapter 2. Other types of subsurface structures must be evaluated using appropriate calculation methods and design standards.

In some (relatively rare) cases, the crane loads and supporting surface strength are all that is needed. This is so when the supporting surface strength is great enough to carry the crane loads without the use of crane mats or pads. Fig. 5.2 shows an outrigger float of a Grove RT600E crane bearing directly on the ground. This setup and job location were such that the 24" (610 mm) diameter floats

Figure 5.2 Grove RT600 Crane without Crane Mats *(David Duerr, P.E.)*

brought the outrigger load down to an acceptably low ground bearing pressure. [For reference, the ground bearing pressure during this lift was about 11,400 psf (546 kPa). Given the nature of the soil (a dry, very well compacted gravel-sand mix with no underground utilities), the ground could safety support this pressure (refer to Table 2.1).]

This is an unusual case, however, so we will now continue with our example crane setup problem in which mats will be required.

Example 5-2 gives the ground conditions for our continuing design example, including the analysis of a subsurface pipe under the crane location.

EXAMPLE 5-2

Soil type – moderately compact sandy clay
Density (γ) – 100 pounds per cubic foot (15.5 kN/m³)
Modulus of soil reaction (E') – 1,000 psi (6,900 kPa)
Water table is below the pipe, so $R_w = 1.0$
Allowable soil bearing pressure (from geotech report) – 3,200 psf (153,000 Pa)
Subsurface pipe – 14" diameter x 0.375" wall (356 mm x 10 mm wall) carbon steel
Pipe $F_y = 36,000$ psi (248,000 kPa)
Cover over pipe (H) – 2.5 feet (0.76 m)
Proposed matting – two layers of 4' x 20' mats, two mats wide (see Fig. 5.7)

The most conservative approach to the analysis of a buried pipe calls for its analysis based on the full allowable soil bearing pressure acting over the full mat area. From Fig. 5.1, the maximum length of the loaded mat area is 296.5", which can be rounded up to 300" = 25'. The maximum possible width of the matted area (pending the mat analysis) is 2 x 20' = 40'. The allowable soil bearing pressure is given as 3,200 psf.

The soil pressure DL_p is calculated with Eq. 2.2.

$$DL_p = H\gamma = 2.5 \times 100 = 250 \text{ psf}$$

The length and width of the bearing area at the top of the pipe are calculated with Eqs. 2.4 and 2.5. The pressure CL_p is then calculated with Eq. 2.7.

$$A = a + 2\frac{H}{\tan 60°} = 25.0 + 2\frac{2.5}{\tan 60°} = 27.89 \text{ feet}$$

$$B = b + 2\frac{H}{\tan 60°} = 40.0 + 2\frac{2.5}{\tan 60°} = 42.89 \text{ feet}$$

$$CL_p = \frac{pab}{AB} = \frac{3,200 \times 25.00 \times 40.00}{27.89 \times 42.89} = 2,676 \text{ psf}$$

Thus, the total pressure P on the plane at the top of the pipe is 250 + 2,676 = 2,926 psf = 20.3 psi.

The ovality of the pipe and the resulting through-thickness bending stress are calculated with Eqs. 2.8 and 2.9. The resulting bending stress is compared to the allowable bending stress of 0.75 F_y = 27,000 psi.

$$I = \frac{0.375^3}{12} = 0.0044 \text{ in.}^4 / \text{ inch of length}$$

$$\frac{\Delta y}{D} = \frac{D_l KP}{\dfrac{(EI)_{eq}}{R^3} + 0.061E'} = \frac{1.5 \times 0.1 \times 20.3}{\dfrac{29{,}000{,}000 \times 0.0044}{7^3} + 0.061 \times 1{,}000} = 0.007$$

$$\sigma_{bw} = 4E\left(\frac{\Delta y}{D}\right)\left(\frac{t}{D}\right) = 4 \times 29{,}000{,}000 \times 0.007 \frac{0.375}{14.000} = 21{,}892 \text{ psi}$$

$$\text{Bending S.R.} = \frac{21{,}892}{27{,}000} = 0.81 < 1.00 \quad \text{O.K.}$$

The bending stress in the pipe wall is found to be acceptable for the defined conditions and crane loading area.

Ring buckling is checked by means of Eqs. 2.10 and 2.11. The ratio of cover H to pipe diameter D is equal to 2.14, so the factor of safety used in Eq. 2.10 is taken as 2.5, as noted in the text following Eqs. 2.10 and 2.11 on page 64.

$$B' = \frac{1}{1+4e^{(-0.065H/D)}} = \frac{1}{1+4e^{(-0.065 \times 30/14)}} = 0.223$$

$$P_c = \frac{1}{FS}\sqrt{32R_w B'E'\frac{(EI)_{eq}}{D^3}} = \frac{1}{2.5}\sqrt{32 \times 1.0 \times 0.223 \times 1{,}000\frac{29{,}000{,}000 \times 0.0044}{14^3}} = 230 \text{ psi}$$

$$\text{Buckling S.R.} = \frac{20.3}{230} = 0.09 < 1.00 \quad \text{O.K.}$$

The critical ring buckling pressure is less than the total applied pressure, so the pipe is again found to be acceptable for the defined conditions and crane loading.

The maximum stress ratio found for the pipe is 0.81 for bending stress. Therefore, the allowable ground bearing pressure q_a is equal to the allowable soil bearing pressure of 3,200 psf (153,000 Pa); q_a is not limited by the strength of the subsurface steel pipe.

The pipe defined in Example 5-2 can be analyzed in any of three ways. The most conservative method, as used in the example, calls for analysis of the pipe loaded by the allowable soil bearing pressure applied over the full loaded mat area. Second, if the first option shows the pipe to be inadequate, the pipe can be analyzed using a reduced allowable ground bearing pressure, where the value is selected to obtain an acceptable result from the pipe analysis. Last, the pipe can be analyzed using the actual applied pressure from the crane acting over the effective mat bearing area determined in the mat design calculations. The method used in Example 5-2 is the desirable approach in that the lift planner is left with only one value, the allowable ground bearing pressure, to use going forward. Of course, if the pipe had been found inadequate using the allowable soil bearing pressure, then analysis based on a reduced allowable ground bearing pressure or the actual crane loading would be necessary.

5.3 CRANE MAT DESIGN – BEARING ON SOIL

The requirements to be met in the design of a crane mat are simple. The mat must have the structural strength and stiffness to distribute the imposed crane load to the supporting surface such that the strength of the supported surface is not exceeded and the supporting surface does not deflect to the point where the crane becomes unacceptably out of level. The calculation methods for accomplishing this goal are developed in Chapter 4, based on the mat strength information in Chapter 3.

The most common types of crane mats are those assembled from rough-sawn timbers. These are also the mats that present the greatest responsibility to the crane user. Appropriate allowable stresses are shown in Table 5.1 for the common (in the U.S.) timber grades, as well as for seven tropical hardwood species used for mat

TABLE 5.1 Suggested Allowable Stresses for Timber Crane Mat Design

(a) Values in U.S. Customary Units:

Species	F_b, psi	F_v, psi	$F_{c\perp}$, psi	E, psi	Density, pcf
Common Species	1,400	200	750	1,200,000	50
Azobe (Ekki)	3,400	410	2,170	1,900,000	60
Dahoma (Dabema)	2,200	220	990	1,300,000	40
Eucalyptus	2,000	265	970	1,300,000	40
Greenheart	3,850	385	1,410	2,200,000	60
Mora	2,500	280	1,090	2,100,000	55
Tonka (Cumaru)	3,850	385	1,650	2,400,000	60
Wamara	4,300	430	2,495	2,200,000	60

(b) Values in SI Units:

Species	F_b, kPa	F_v, kPa	$F_{c\perp}$, kPa	E, kPa	Density, kg/m³
Common Species	9,650	1,380	5,150	8,300,000	800
Azobe (Ekki)	23,500	2,820	15,000	13,100,000	960
Dahoma (Debema)	15,300	1,530	6,800	9,000,000	640
Eucalyptus	13,600	1,840	6,600	9,000,000	640
Greenheart	26,500	2,650	9,800	15,200,000	960
Mora	17,300	1,940	7,400	14,500,000	880
Tonka (Cumaru)	26,500	2,650	11,400	16,600,000	960
Wamara	29,600	2,960	17,200	15,200,000	960

construction. Fabricated steel mats are generally designed for a specified load and proprietary mat products, such as the synthetic mats discussed in Chapters 3 and 4, have rated loads or other detailed use guidance from their manufacturers. Timber mats, however, must be fully evaluated by the crane user or lift planner to assure their adequacy for a given crane installation.

5.3.1   Crane Mats Under Outriggers

The design of a crane mat to support an outrigger is usually very simple. The mat is centered under the outrigger, if possible. If this is not possible, then the mat is treated as symmetrical and of a length equal to twice the shortest distance from the center of the outrigger to the end of the mat. The excess mat on the long side is ignored.

When sizing a mat for use under an outrigger, consideration must be given to the size of the outrigger float. For example, an 18" (460 mm) diameter float will bear on only two timbers of a mat made up of 12" x 12" (300 mm x 300 mm) timbers and the tie rods that hold the mat together are not necessarily capable of distributing this concentrated load to the other timbers. In this case, a conservative approach calls for checking the mat considering only the two timbers on which the pad bears. Alternately, a steel plate, a synthetic pad, or timbers placed crosswise can be used to distribute the outrigger load to all of the mat timbers (Fig. 5.3). This practice is discussed further in Section 4.4.5.

Just as one long mat can be loaded by both tracks of a crawler crane, two outriggers can bear on one mat. Such an arrangement is usually only practical with a smaller crane where the outrigger spacing is no more than 20 to 22 feet (6 to 7 meters). The analysis approach shown in Section 4.4.4 is appropriate can be applied to this case, as well.

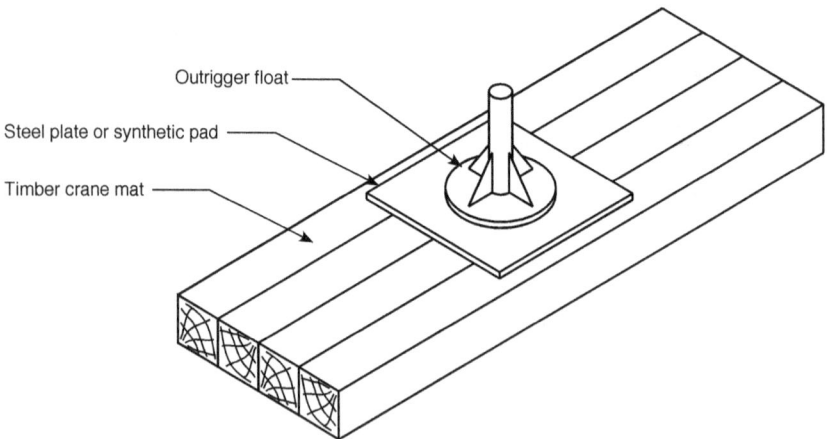

Outrigger float

Steel plate or synthetic pad

Timber crane mat

Figure 5.3 Steel Plate or Synthetic Pad over Timber Mat

5.3.2   Crane Mats Under Crawler Tracks

The most common mat arrangements under a crawler crane is a single layer of long mats under both tracks or two runs of mats side by side, with one track bearing on each mat run. When using timber mats, one mat under both tracks is generally only practical for small to medium cranes, as timber mats are rarely available at lengths above about 30 feet (9 meters). The track spread on larger crawler cranes necessitates the two runs of mats. The magnitude of the track pressures and/or the strength (or lack thereof) of the soil may require two or more layers of mats, either parallel or perpendicular to each other. These various mat configurations are discussed in detail in Section 4.4.

The track bearing pressure applied by the crawler track to the mats typically varies from one end to the other, either in a triangular pattern or a trapezoidal pattern. A long-standing practice (Shapiro and Shapiro 2011, including earlier editions) has been to approximate the pressure distribution to mats that are perpendicular to the track as a step function by using the average bearing pressure over the width of the mat, most commonly 4 feet (1.2 meters). This approach avoids having a mat design that is based on the peak pressure at the end of the track and has been felt to be a more realistic model of the interaction of the crawler track, the mats, and the soil. (It is noted that an eccentrically loaded concrete footing is typically designed using the peak pressure between the footing and the soil. It is also noted that a concrete footing is significantly more rigid than crane mats.)

Consider the mat-supported crawler track in Fig. 5.4a. The track bearing pressure envelope is illustrated in Fig. 5.4b, showing a peak bearing pressure at the right end of the rollers, tapering down to a lower pressure at the left end. Fig. 5.4c

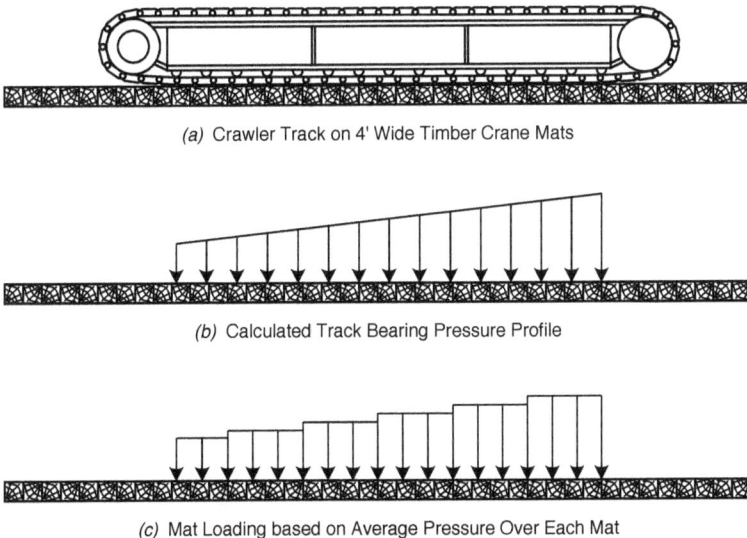

(a) Crawler Track on 4' Wide Timber Crane Mats

(b) Calculated Track Bearing Pressure Profile

(c) Mat Loading based on Average Pressure Over Each Mat

Figure 5.4 Track Pressure Distribution to Crane Mats

shows this pressure diagram converted to average pressures over each 4-foot wide mat. It is this average track bearing pressure that is used for the design of the mats that are perpendicular to the length of the track.

In addition to this practice of averaging the pressure over the width of a mat, the author has also seen suggestions of using a pressure averaged over both shorter and longer fractions of the track bearing length. The use of such an approximation raises two obvious questions:

1. How much of an error is introduced by this approach?
2. How long an area may be used to average the track bearing pressure?

Both of these questions can be answered by running through calculations of support reactions using both the actual track bearing pressure diagram (triangular or trapezoidal in shape) and the corresponding loads to each mat, using mats of varying widths and assuming that the load to each mat is based on the average pressure (Fig. 5.4c). We can look at the results of these calculations in either of two ways. First, we can simply look at the peak bearing pressure from the actual bearing pressure diagram compared to the average pressure on the most heavily loaded mat. Second, the result of each calculation can expressed as the moment due to the eccentricity of the centroid of the actual pressure diagram (e.g. Fig. 5.4b) to the centroid of the stepped pressure diagram (e.g. Fig. 5.4c). This second effect is illustrated in Fig. 5.5.

The error introduced by this bearing pressure approximation is either the difference in maximum bearing pressure in the first comparison or the difference in centroid position, which can be normalized as follows. A moment equal to the actual applied load acting through its centroid times the distance from the center of the track bearing length to the centroid is calculated. A corresponding moment is calculated for the stepped bearing pressure diagram. The error is calculated as the difference between these two moments divided by the moment based on the actual track bearing pressure diagram.

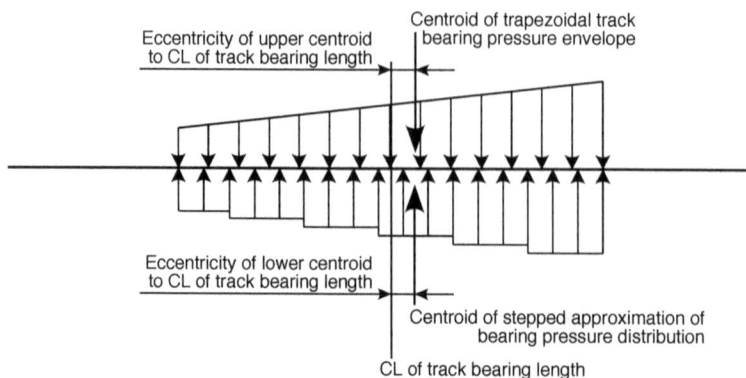

Figure 5.5 Relationship of Track Bearing Pressure Centroids

The results of this approximation, using average bearing widths ranging from zero (no averaging of the track bearing pressure) to 50% of the track bearing length are illustrated in Fig. 5.6. The straight line plot of the track bearing pressure deviation is based on a triangular pressure diagram; the maximum value of the bearing pressure deviation is lower for trapezoidal diagrams, but the deviation still plots as a straight line. The curved plot for deviation in centroid position is the same for all bearing pressure envelopes.

As an example, a 4-foot wide mat is 20% of the track bearing length for a 20-foot long track. This equates to an error of about 4% in the overturning moment due to the shift in the location of the centroid of the supported load and as much as 10% in the calculation of maximum bearing pressure. This is not significant with respect to the overall accuracy with which these values are known. On the other hand, averaging the track bearing pressure over one-third of the bearing length, the error introduced is about 11% of the overturning moment due to the shift in the centroid position or 17% of the maximum bearing pressure, which may be considered to be unacceptable. The worst case of dividing the track bearing length into two halves (that is, the pressure is averaged over 50% of the bearing length) equates to an error of 25% for both comparisons, which is clearly not acceptable.

With this basic information in hand, we can continue with our example crane support design problem. Two mat arrangements will be analyzed. Example 5-3 illustrates the analysis of two layers of mats in which both layers are perpendicular to the long dimension of the tracks (Fig. 5.7). Example 5-4 illustrates the analysis of two layers of mats in which the top mat is parallel to the long dimension of the track and the bottom layer is perpendicular (Fig. 5.8). The examples demonstrate the use of the approximation of averaging the track bearing pressure over the width of one mat.

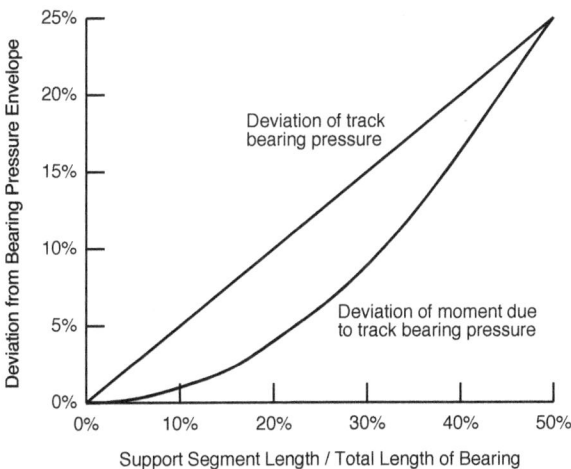

Figure 5.6 Variation of Error Introduced by Bearing Width Approximation

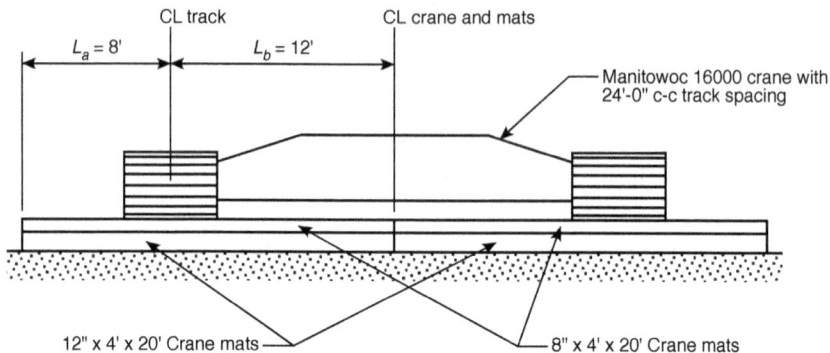

Figure 5.7 Crane Mat Arrangement for Example 5-3

EXAMPLE 5-3

Layout of standard timber crane mats as shown in Fig. 5.7:

12" x 4' x 20' lower mats and 8" x 4' x 20' upper mats

Track bearing pressures from Example 5-1, shown in Fig. 5.1b (swinging with no hook load governs the installation design)

Allowable ground bearing pressure q_a = 3,200 psf (153,000 Pa) from Example 5-2

From Fig. 5.1b, we find a peak track bearing pressure of 81.8 psi, dropping to 0.0 psi over a length of 237.3 inches. Thus, the rate of change of bearing pressure is (81.8 - 0.0) / 237.3 = 0.345 psi per inch, the bearing pressure at the inside edge of the most heavily loaded mat is 81.8 - 0.345 x 48 = 65.3 psi, and the total load to the mat is as follows, based on a track hard bearing width of 50" and a mat width of 48".

$$P = 50 \times 48 \times \frac{81.8 + 65.3}{2} = 176,465 \text{ pounds}$$

The shorter distance from the center of a crawler track to the end of a mat is shown in Fig. 5.7 as 8'-0". Thus, the symmetrical length of the mat is 16'-0". This is the maximum bearing length of the mat. The effective bearing length is now calculated using the method developed in Chapter 4.

We now apply Eq. 4.39 to determine how the applied load is distributed between the 8" top mats and the 12" bottom mats. Since the modulus of elasticity E of both sizes of timbers has the same value, we can simplify the equation by reducing it to just the ratio of the moments of inertia, as shown below.

$$L_x = L_t \frac{E_x I_x}{\sum_{i=1}^{i=n} E_i I_i} \Rightarrow L_{12} = L_t \frac{E_{12} I_{12}}{E_8 I_8 + E_{12} I_{12}} = L_t \frac{6,912}{2,048 + 6,912} = 0.7714 L_t$$

$$L_x = L_t \frac{E_x I_x}{\sum_{i=1}^{i=n} E_i I_i} \Rightarrow L_8 = L_t \frac{E_8 I_8}{E_8 I_8 + E_{12} I_{12}} = L_t \frac{2,048}{2,048 + 6,912} = 0.2286 L_t$$

From this calculation, we see that 22.86% of the load will be carried by the 8" mats and 77.14% of the load will be carried by the 12" mats. We next calculate the bending and shear strengths of both sizes of mats using Eq. 4.21 for bending and Eq. 4.24 for shear.

$$M_n = F_b S = 1,400 \frac{12^2 \times 48}{6} = 1,612,800 \text{ pound-inches} = 134,400 \text{ pound-feet}$$

$$V_n = F_v \frac{Bd}{1.5} = 200 \frac{48 \times 12}{1.5} = 76,800 \text{ pounds}$$

$$M_n = F_b S = 1,400 \frac{8^2 48}{6} = 716,800 \text{ pound-inches} = 59,733 \text{ pound-feet}$$

$$V_n = F_v \frac{Bd}{1.5} = 200 \frac{48 \times 8}{1.5} = 51,200 \text{ pounds}$$

Based on these results, the 12" mats have $134,400 / (134,400 + 59,733) = 69.23\%$ of the bending strength of the combined mats and $76,800 / (76,800 + 51,200) = 60.00\%$ of the shear strength of the combined mats, but will carry 77.14% of the total load. Since the share of the load carried by the 12" mats is greater than their share of the strength, the effective bearing length calculation is to be based on the behavior of the 12" mats.

Dead weight of the 12" mats $W = 1' \times 4' \times 20' \times 50 \text{ pcf} = 4,000 \text{ pounds}$
Dead weight of the 8" mats $W = 0.67 \times 4' \times 20' \times 50 \text{ pcf} = 2,667 \text{ pounds}$
The effective bearing length of the mat is now calculated using Eqs. 4.19 through 4.31, as applicable. Note the use of the values of pct in these calculations.
Calculate the effective bearing length L_{eff} as limited by bending strength.

$$L_{max} = 0.7714 L_t \Rightarrow pct = \frac{L_{max}}{L_t} = \frac{0.7714 L_t}{L_t} = 0.7714$$

$$\left(q_a B\right) L_{eff}^2 + \left(-2q_a BC - W\right) L_{eff} + \left(q_a BC^2 + 2CW - 8M_n / pct\right) = 0$$

$$\left(3,200 \times 4.00\right) L_{eff}^2 + \left[-2 \times 3,200 \times 4.00 \times 4.17 - \left(4,000 + 2,667\right)\right] L_{eff}$$

$$+ \left[3,200 \times 4.00 \times 4.17^2 + 2 \times 4.17 \times \left(4,000 + 2,667\right) - 8 \times 134,400 / 0.7714\right] = 0$$

$$\left(12,800\right) L_{eff}^2 + \left(-113,333\right) L_{eff} + \left(-1,116,000\right) = 0$$

$$L_{eff} = \frac{-\left(-113,333\right) \pm \sqrt{\left(-113,333\right)^2 - 4\left(12,800\right)\left(-1,116,000\right)}}{2\left(12,800\right)} = 14.761 < L_{mat} = 16.000$$

Calculate the effective bearing length L_{eff} as limited by shear strength.

$$\left(q_a B\right) L_{eff}^2 + \left(-2V_n / pct - q_a BC - 2q_a Bd - W\right) L_{eff} + \left(WC + 2Wd\right) = 0$$

$$\left(3,200 \times 4.00\right)L_{eff}^2$$

$$+\left[-2 \times 76,800/0.7714 - 3,200 \times 4.00 \times 4.17 - 2 \times 3,200 \times 4.00 \times 1.00 - \left(4,000+2,667\right)\right]L_{eff}$$

$$+\left[\left(4,000+2,667\right) \times 4.17 + 2 \times \left(4,000+2,667\right) \times 1.00\right] = 0$$

$$\left(12,800\right)L_{eff}^2 + \left(-284,711\right)L_{eff} + \left(41,111\right) = 0$$

$$L_{eff} = \frac{-\left(-284,711\right) \pm \sqrt{\left(-284,711\right)^2 - 4\left(12,800\right)\left(41,111\right)}}{2\left(12,800\right)} = 22.098 > L_{mat} = 16.000$$

Calculate the effective bearing length L_{eff} as limited by deflection.

$$L_c = 3\sqrt{\frac{0.06\sum\left(EI\right)}{0.9q_aB}} = 3\sqrt{\frac{0.06\left(1,200,000\right)\left(2,048+6,912\right)}{0.9\left(3,200/144\right)48.000}} = 87.590 \text{ inches} = 7.299 \text{ feet}$$

$$L_{eff} = 2L_c + C = 2 \times 7.299 + 4.17 = 18.765 > L_{mat} = 16.000$$

The smallest value of L_{eff} calculated above is 14.761 feet, as limited by bending strength. Thus, this value is the effective bearing length of the mat for this application The mat forces and ground bearing pressure are now calculated as follows, using Eqs. 4.32 through 4.36. The calculated bending moment, shear, and ground bearing pressure are expressed relative to the allowable values as utilization ratios. Although we know that the strength of the 12" mats governs the design, the 8" mats should be checked as well.

$$q = \frac{P}{L_{eff}B} = \frac{176,465}{14.761 \times 4.00} = 2,989 \text{ psf}$$

$$L_c = \frac{L_{eff} - C}{2} = \frac{14.761 - 4.167}{2} = 5.297 \text{ feet}$$

$$M = \frac{\left(pct \ q \ B\right)L_c^2}{2} = \frac{\left(0.7714 \times 2,989 \times 4.00\right)5.297^2}{2} = 129,386 \text{ pound-feet}$$

$$\text{Bending Stress Ratio} = \frac{M}{M_n} = \frac{129,386}{134,400} = 0.963$$

$$V = pct \ q \ B\left(L_c - d\right) = 0.7714 \times 2,989 \times 4.00\left(5.297 - 1.00\right) = 39,629 \text{ pounds}$$

$$\text{Shear Stress Ratio} = \frac{V}{V_n} = \frac{39,629}{76,800} = 0.516$$

$$q_t = \frac{P+W}{L_{eff}B} = \frac{176,465 + \left(4,000+2,667\right)}{14.761 \times 4.00} = 3,102 \text{ psf}$$

$$\text{Ground Bearing Pressure Ratio} = \frac{q_t}{q_a} = \frac{3,102}{3,200} = 0.969$$

$$M = \frac{(pct\ q\ B)L_c^2}{2} = \frac{(0.2286 \times 2,989 \times 4.00)5.297^2}{2} = 38,337 \text{ pound-feet}$$

$$\text{Bending Stress Ratio} = \frac{M}{M_n} = \frac{38,337}{59,733} = 0.642$$

$$V = pct\ q\ B(L_c - d) = 0.2286 \times 2,989 \times 4.00(5.297 - 0.667) = 12,653 \text{ pounds}$$

$$\text{Shear Stress Ratio} = \frac{V}{V_n} = \frac{12,653}{51,200} = 0.247$$

The maximum utilization ratio is found to be about 0.96 for both the strength of the 12" mats and the ground bearing pressure, which is acceptable. (The small difference between the bending stress ratio and the ground bearing pressure ratio is due to the approximation introduced in the derivation of Eq. 4.19, as is discussed in the accompanying text in Section 4.2.1.)

There is one last calculation that will be performed as a part of this example, although it is not normally needed in practice. That is a calculation of the deflection of each mat layer. The mats obviously will remain in contact with one another, which means that they exhibit identical deflection. Calculation of these deflections validates this analysis method.

$$\Delta_{12} = \frac{(pct\ w)L_c^4}{8EI} = \frac{\left[0.7714\left(\frac{2,989 \times 4}{12}\right)\right](5.297 \times 12)^4}{8(1,200,000)\frac{12^3 \times 48}{12}} = 0.189 \text{ inch}$$

$$\Delta_8 = \frac{(pct\ w)L_c^4}{8EI} = \frac{\left[0.2286\left(\frac{2,989 \times 4}{12}\right)\right](5.297 \times 12)^4}{8(1,200,000)\frac{8^3 \times 48}{12}} = 0.189 \text{ inch}$$

EXAMPLE 5-4

Layout of standard timber crane mats as shown in Fig. 5.8:

12" x 4' x 20' lower mats arranged similarly to the layout shown in Fig. 5.7 and one 12" x 5' x 32' upper mat centered in both directions under each crawler track

Track bearing pressures from Example 5-1, shown in Fig. 5.1b (swinging with no hook load governs the installation design)

Allowable ground bearing pressure q_a = 3,200 psf (153,0000 Pa) from Example 5-2

The cantilevered length of the upper mat L_c is initially taken as the actual distance from the last roller to the end of the mat. It is found that the upper mat is overstressed with this value of L_c. Through trial and error, we converge on a value of L_c of 32.50 inches. Using L_c = 32.50

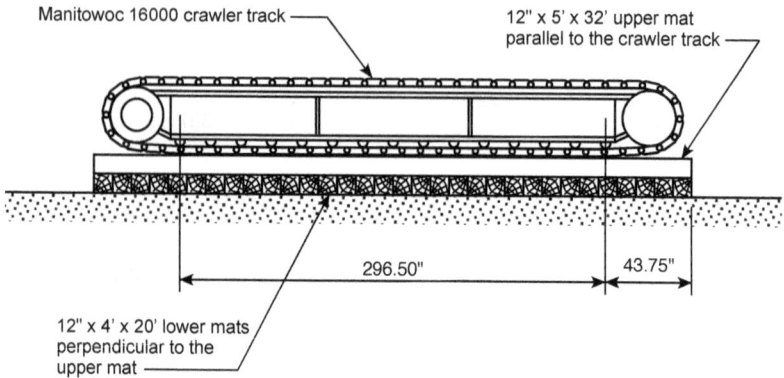

Figure 5.8 Crane Mat Arrangement for Example 5-4

inches, the other dimensions shown in Fig. 5.8, and the track bearing pressures of the more heavily loaded track when the crane swings with no load on the hook (Fig. 5.1b), calculate the crawler track load P, the position the centroid of P from the more heavily loaded end of the track c', and the eccentricity of the centroid of P with respect to the center of the crawler track e_t using Eqs. 4.40, 4.41, and 4.42, respectively.

$$P = \frac{(t_{max} + t_{min})}{2} CL = \frac{(81.80 + 0.00)}{2} 50.00(237.30) = 485,279 \text{ pounds}$$

$$c' = \frac{3Lt_{min} + L(t_{max} - t_{min})}{3(t_{max} + t_{min})} = \frac{3(237.30)0.00 + 237.30(81.80 - 0.00)}{3(81.80 + 0.00)} = 79.10 \text{ inches}$$

$$e_t = \frac{L}{2} - c' = \frac{237.30}{2} - 79.10 = 39.55 \text{ inches}$$

Dimensions and results are illustrated in Fig. 5.9.

Since the centroid of the track load P on the top of the mat must be coaxial with the

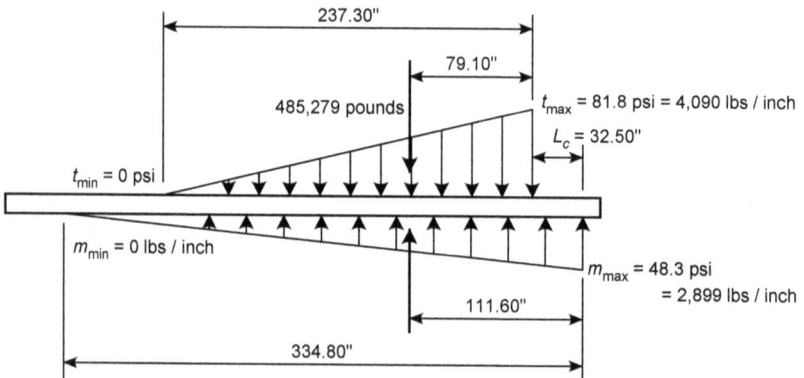

Figure 5.9 Bearing Pressure Distribution Through Upper Mat

centroid of the bearing pressure envelope on the bottom of the mat, the lower centroid position is calculated as follows.

$$c' + L_c = 79.10 + 32.50 = 111.60 \text{ inches}$$

We can now calculate the effective bearing length L_{eff} of the upper mat, the eccentricity e_m of the centroid of the load P relative to the center of L_{eff}, and the maximum and minimum bearing pressures m_{max} and m_{min} imposed by the upper mat onto the lower mats using Eqs. 4.43, 4.44, 4.45, and 4.46, respectively.

$$L_{eff} = 3(c' + L_c) \leq L_{mat} = 3(111.60) \leq 384.00 = 334.80 \text{ inches}$$

$$e_m = \frac{L_{eff}}{2} - (c' + L_c) = \frac{334.80}{2} - 111.60 = 55.80 \text{ inches}$$

$$m_{max} = \frac{P}{L_{eff}}\left(1 + \frac{6e_m}{L_{eff}}\right) = \frac{485,279}{334.80}\left[1 + \frac{6(55.80)}{334.80}\right] = 2,899 \text{ pounds per inch}$$

$$m_{min} = \frac{P}{L_{eff}}\left(1 - \frac{6e_m}{L_{eff}}\right) = \frac{485,279}{334.80}\left[1 - \frac{6(55.80)}{334.80}\right] = 0 \text{ pounds per inch}$$

These bearing pressures acting on the bottom surface of the upper mat are now used to calculate the bending, shear, and deflection of the mat.

Calculate the bearing pressure in line with the last track roller m_{roller}, the total load acting on the cantilevered segment of the upper mat P_c, the centroid of that total load c, and the resulting moment and bending stress.

$$m_{roller} = m_{max} - \frac{(m_{max} - m_{min})}{L_{eff}}L_c = 2,899 - \frac{(2,899 - 0)}{334.80}32.50 = 2,618 \text{ pounds per inch}$$

$$P_c = \frac{m_{max} + m_{roller}}{2}L_c = \frac{2,899 + 2,618}{2}32.50 = 89,642 \text{ pounds}$$

$$c = \frac{L_c(2m_{max} + m_{roller})}{3(m_{max} + m_{roller})} = \frac{32.50[2(2,899) + 2,618]}{3(2,899 + 2,618)} = 16.53 \text{ inches}$$

$$M = P_c c = 89,642(16.53) = 1,481,451 \text{ pound-inches}$$

$$f_b = \frac{M}{S_x} = \frac{1,481,451}{12^2 60/6} = 1,029 \text{ psi}$$

$$\text{Bending S.R.} = \frac{f_b}{F_b} = \frac{1,029}{1,400} = 0.73$$

The same calculation procedure is used to calculate the bearing pressure at a distance from the last roller equal to the thickness of the mat d (which is the point at which shear is evaluated in a timber mat) m_{shear}, the shear force, and the shear stress.

$$m_{shear} = m_{max} - \frac{(m_{max} - m_{min})}{L_{eff}}(L_c - d) = 2,899 - \frac{(2,899 - 0)}{334.80}(32.50 - 12.00) = 2,721 \text{ lbs / inch}$$

$$V = \frac{m_{max} + m_{shear}}{2}(L_c - d) = \frac{2,899 + 2,721}{2}(32.50 - 12.00) = 57,608 \text{ pounds}$$

$$f_v = \frac{1.5V}{A} = \frac{1.5(57,608)}{12(60)} = 120 \text{ psi}$$

$$\text{Shear S.R.} = \frac{f_v}{F_v} = \frac{120}{200} = 0.60$$

Last, the deflection of the upper mat is calculated. Here, a conservative approximation can be used to simplify the calculations. Rather than accounting for the changing bearing pressure, the deflection can be calculated using the maximum bearing pressure acting along the full cantilevered length L_c. If a more precise solution is needed, Case 2a, Table 8.1 in Young, et al (2012) can be used.

$$\Delta = \frac{m_{max}L_c^4}{8EI} = \frac{2,899(32.50^4)}{8(1,200,000)12^3 60/12} = 0.039 \text{ inch}$$

$$\text{Deflection Ratio} = \frac{0.039}{32.50} = 0.0012 = 0.12\% \text{ of } L_c$$

We find that the bending and shear stresses are acceptable and the deflection is very small. Thus, the specified upper mat is effective for longitudinal load spreading at the iteratively determined value of L_c.

The lower mats are now analyzed in the normal manner using the pressures and length of bearing determined here as acting between the upper mat and lower mats. Note that the width of bearing dimension C used in the lower mat calculations is the width of the upper mat (60 inches in this example).

It is also noted that this mat arrangement does not yield a closed balanced design. In this example, the effective value of L_c was determined iteratively (that is, by trial and error) until a reasonably balanced solution was developed. Here, we have found that the maximum stress ratio in the upper mat is shown to be 0.73 (bending stress). Continuation of the design problem shows that the maximum stress ratio in the lower mats is 0.73 (bending stress) and the ground bearing pressure utilization ratio is 0.74. If the upper mat stress ratio was smaller than the lower mat or ground bearing ratios when using the actual value of L_c (43.75", as shown in Fig. 5.8), then the upper mat would have been acceptable over its full length, no adjustment of L_c would be required, and the results of the upper mat analysis would be used for the check of the lower mats. A reduction of L_c may also be necessary if the mat deflection is excessive.

We have now stepped through a mobile crane installation design using the methods developed in this handbook.   The purpose of this exercise is the demonstration of the process of designing crane supports.   As noted at the beginning of this chapter, it is not the intention to present this material as a crane support design cookbook.   Although the basic process remains fairly constant from one crane setup to the next, there are differences, sometimes very significant differences, that the lift planner/engineer must account for in the support design. Blindly plugging numbers into equations isn't good enough.   The better one understands the material covered in Chapters 1 through 4, the better prepared one will be to tackle the unique problems that arise periodically.   Only through such a true understanding of the principles at work can safety and efficiency be optimized.

5.4   CRANE MAT DESIGN – BEARING ON STRUCTURES OR PAVEMENT

This section considers the situation in which the crane mats bear directly on a structure, rather than on soil.   This occurs when a crane is set up on the floor framing of a building, on a bridge, on the deck of a barge, on a large foundation, or on paving, such as a road, a parking lot or a building floor that is a slab on grade. The more technical aspects of this installation design are discussed in Section 2.5. A list of design references applicable to the various types of structures discussed is provided in Section 2.5.3.   As noted in the introductory paragraphs of Section 2.5, due to the broad range of structures that may be encountered in mobile crane installation work, the guidance given in this handbook is relatively general. The lift planner is often well advised to seek out additional expertise, such as a structural engineer or a naval architect.

5.4.1   Crane Mats on Slabs or Pavement

The need to set up a mobile crane on a paved area, such as a street, parking lot or any other area that is covered by a concrete slab on grade, often occurs when working in an urban job site.   In these situations, the crane mat design equations developed in Chapter 4 are generally not applicable.   Other approaches must be employed to determine the bearing area of the crane mats.

When setting up a small crane or utility truck on asphalt pavement, sometimes all that is needed is a plate that can spread the outrigger load out enough to avoid damaging the surface of the pavement. Fig. 5.10 shows the use of a relatively thin (0.50"; 13 mm) aluminum plate under an outrigger float on the asphalt paving of a parking lot.

The mat design method of Chapter 4 depends on the compressibility of the underlying soil to allow the mat to spread the supported crane load. The stiffness presented by pavement will not permit this same level of deformation.   Thus,

Figure 5.10 Aluminum Plate Under a Lightly Loaded Outrigger *(David Duerr, P.E.)*

relatively little load spreading will be provided by typical crane mats placed on a paved area. Rigorous analysis of the stiffness of a paved area is often not practical, usually due to the lack of detailed information about the pavement and subgrade preparation. The following approximations are suggested as reasonable starting points. Deviations from these suggestions may be appropriate in some cases and must be supported by sound engineering judgment.

When crane mats are laid over a rigid pavement (e.g. a reinforced concrete slab on grade), the load spreading provided by the mats may reasonably be considered to be equal to the thickness of the mat on each side of the crane's bearing surface (track or outrigger float). That is, the load spreads on a 1 : 1 slope through the mat material. As an example, if a crawler crane with a track pad bearing width of 30" (762 mm) is set up on 12" (305 mm) thick mats over a reinforced concrete slab on grade, the width of bearing to the slab may be taken as 12" + 30" + 12" = 54" (305 mm + 762 mm + 305 mm = 1,372 mm).

When crane mats are laid over a flexible pavement (e.g. bituminous asphalt), the load spreading provided by the mats will be somewhat greater than that which occurs with the rigid pavement due to the lower stiffness of the flexible pavement. Here, the slope of the load spreading through the mats may be taken as equal to a slope of at least 1.5 : 1. Repeating the previous example, if a crawler crane with a track pad bearing width of 30" (762 mm) is set up on 12" (305 mm) thick mats over asphalt pavement, the load spreading on each side is equal to 1.5 x 12 = 18" (457 mm) and the width of bearing to the pavement may be taken as 18" + 30" + 18" = 66" (457 mm + 762 mm + 457 mm = 1,676 mm).

These approximations are generally suitable for use with any type of crane mat (i.e. timber, synthetic, or steel). If special load spreading elements are designed with a particularly high bending stiffness, then greater load spreading may be obtained. This must be evaluated on a case by case basis.

5.4.2   Crane Mats on Structures

Mobile cranes are often set up on a structure. This may be the floor or roof of a building, a bridge, or a barge deck (see Section 2.5.2). Understandably, the interaction between the crane, the crane mats, and the supporting structure is significantly different than the interaction with soil.

Consider the crawler track illustrated in Fig. 5.11. The upper figure shows a typical crawler track bearing pressure diagram based on bearing on a hard, flat surface. When the crawler is supported on a framed structure, the support provided by that structure will not be uniform. The crawler track bearing pressure must be resolved into a set of concentrated loads at the larger framing members, as illustrated in the lower figure. The use of a longitudinal crane mat to spread the load out to the members beyond the bearing length of the track is possible, but it's success is dependent upon the mat having a bending stiffness that is significant relative to the vertical stiffness of the structure's hard points.

In the case of a building or bridge structure with a substantial deck, the deck may also provide load spreading. In the case of a flat deck steel barge, the deck structure is often relatively light, so the crawler load will be carried only by those structural elements, such as internal bulkheads or trusses, that are directly below the crawler unless very stiff mats are employed.

The design of the support of outrigger loads on a framed structure (Fig. 5.12) is generally simpler than support design for a crawler track. In this arrangement, the crane mat serves as a simple beam between hard points to support the outrigger. The most conservative approach calls for assuming that only the hard points on either side of the outrigger serve to provide support to the mat. Unless the mat is

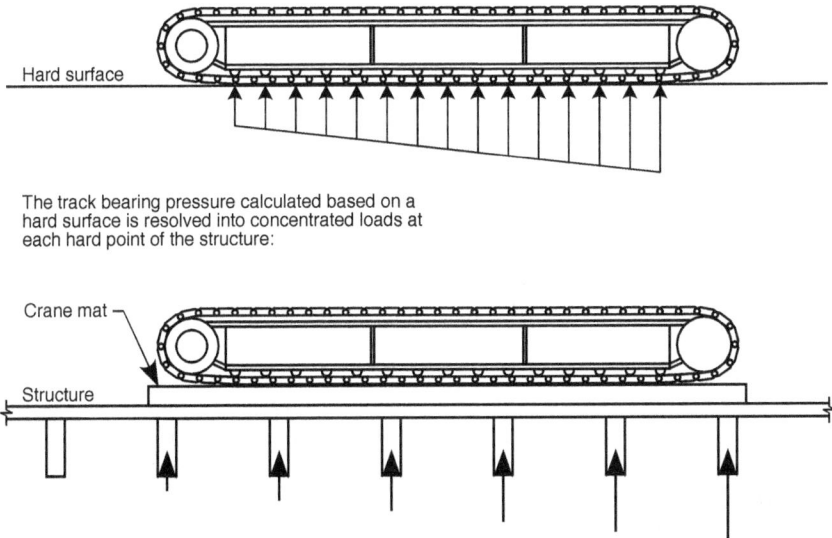

Hard surface

The track bearing pressure calculated based on a hard surface is resolved into concentrated loads at each hard point of the structure:

Crane mat

Structure

Figure 5.11 Crawler Crane Supported on a Framed Structure

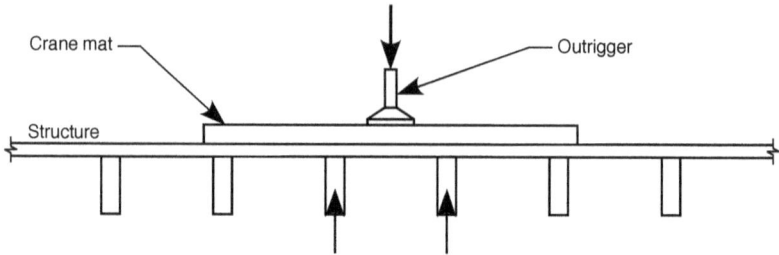

Figure 5.12 Outrigger Supported on a Framed Structure

very stiff in bending relative to the vertical stiffness of the framing elements, this is a reasonably accurate representation of the behavior that will occur.

We can see from Fig. 5.12 that placing an outrigger directly over a single framing member, such as a floor beam in a building or a bulkhead in a barge, can present a load distribution problem. In this case, it is likely that the full outrigger load will be carried by the single hard point. If the beam, bulkhead, or other element is capable of carrying the load, then the setup is acceptable. For larger cranes in particular, this is often not the case, so positioning the crane so the outriggers fall as shown in Fig. 5.12 is usually desirable.

Also of importance is the determination of any other loads and forces that will act on the structure in combination with the loads from the mobile crane. For example, if a crane is set up on a ground level floor structure that spans over a basement, the floor may be supporting other loads in addition to the crane. There may be other pieces of equipment on the floor or equipment, piping or ductwork suspended from the underside of the floor. All of these loads and forces must be determined to assure that the evaluation of the structure is based on realistic loads and load combinations.

5.4.3   Applicable Standards

Suggested references that are useful for the evaluation of various types of structures are listed in Section 2.5.3 are repeated here.

- Steel building structures are typically analyzed using the AISC *Specification for Structural Steel Buildings* (AISC 2016). Additional guidance on load factors and load combinations for relatively short-term conditions can be found in ASCE (2015).
- Concrete building structures, including many foundation elements, are designed or analyzed using the ACI *Building Code Requirements for Structural Concrete* (ACI 2014). Again, guidance on load factors and load combinations can be found in ASCE (2015).
- Bridge structures are analyzed using the *AASHTO LRFD Bridge Design Specifications* (AASHTO 2017).

- Flexible pavements can be analyzed using the method detailed in MS-1 *Thickness Design – Highways & Streets* (Asphalt Institute 1999). Some interpretation will be necessary to apply this standard to the large area of a mobile crane support since this standard is based on loads being applied to the pavement by wheeled vehicles.
- A concrete slab on grade, which can be pavement or a building floor, can be analyzed using the procedures in *Concrete Floors on Ground* (Farny and Tarr 2008). As with the analysis of flexible pavement, some interpretation and extrapolation will be necessary since this standard is also based on loads applied to the surface by wheeled vehicles, as well as from posts.
- Barge structures are proportioned using the *Rules for Building and Classing Steel Barges* (ABS 2017). Part 5, Chapter 3, Section 3 of this document specifically addresses the design of crane barges.
- Most cities throughout the U.S. have local building codes that may have requirements that go beyond those established in the industry standards listed above. The contractor must check with local building officials to assure that any such requirements are met. Also, the owner of the facility where the work is being performed can often provide guidance with respect to applicable local codes.

Some of these structures can be evaluated by many lift planning specialist engineers who have an appropriate background in civil engineering. However, as previously noted, the evaluation of complex or specialized structures may require the involvement of a structural engineer or naval architect.

5.5   CRANE SUPPORT WHILE TRAVELING

We most commonly think about the support of a crane while it is in one location. This obviously is almost always the case for outrigger supported cranes (although few and far apart, there are some truck cranes that have been fitted with rolling outriggers). Crawler cranes, however, quite often travel. This may be just for positioning the crane or the crane may travel with a load on the hook. In either case, the support design must account for the changing conditions as the crane moves.

If the crane is supported solely by mats oriented perpendicular to the crawler tracks, as illustrated in Figs. 5.4 and 5.7, then no particular work is needed to accommodate travel, other than assuring that the allowable ground bearing pressure is adequate over the full travel path. If the mat arrangement utilizes an upper layer of mats running parallel to the tracks, then the ability of the arrangement to support the crane when the load spreading of the top layer is lost (see Section 4.4.2) must be analyzed.

There is one more case that has not yet been discussed here. What happens when the crane's travel path takes it from one type of surface to another where the two surfaces have significantly different stiffnesses? Consider the situation illustrated in Fig. 5.13. A crawler crane is set up on a substantial concrete foundation and must travel off the foundation onto the adjacent soil. The heavy concrete foundation is, for all practical purposes, unmoving under the load of the crane. The soil, on the other hand, will compress. The support of the tracks can be likened to the support provided by two groups of springs of different stiffness (Fig. 5.14).

If the two surfaces are both yielding, just with different stiffnesses, the condition will be as shown in Fig. 5.14*a*. That is, the two surfaces can be represented by two sets of springs with different spring constants. If one surface is effectively unyielding, as we described the concrete foundation illustrated in Fig. 5.13, then the support of the tracks is more like Fig. 5.14*b*. In this case, the combination of the hard point support and the compressing soil will allow the crane to go out of level. If the soil is soft enough or if the crane's center of gravity position is far enough toward the end of the crane on the soil, the crane can overturn.

Whether traveling with a load or simply to reposition the crane for the next lift, we can see that consideration of the nature of the travel path is of great significance. This is particularly true when repositioning the crane with no load on the hook. Moving the crane without a load on the hook is all too often considered to be "routine" or "unimportant." It is not. An appropriate level of care must be taken during all crane operations.

5.6   RULES OF THUMB – DO THEY WORK?

When the subject of crane support comes up in lift planning work, almost without fail someone will suggest using a shortcut or rule of thumb to evaluate the support requirements. The first edition of this handbook avoided discussion of these rules of thumb, simply because they are rarely, if ever, valid. However, reader feedback has driven the inclusion of the following discussion in this second edition.

This section examines a few of the more common crane support rules of thumb, if for no other reason than to highlight their shortcomings.

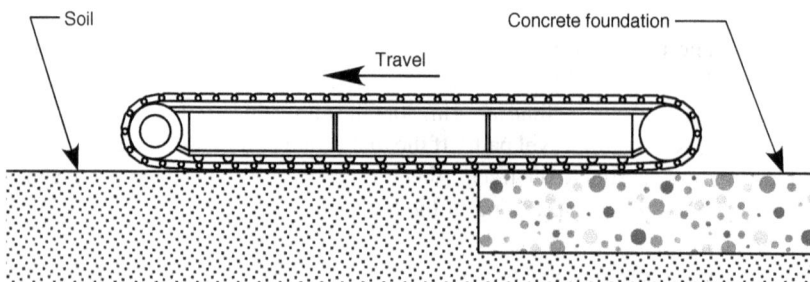

Figure 5.13 Crawler Crane Transitioning Between Surfaces

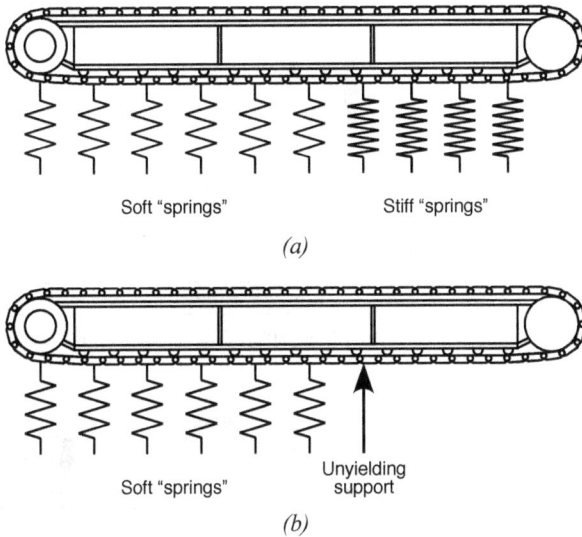

Figure 5.14 Track Support Represented as Springs

5.6.1 Crawler Cranes

The most common rule of thumb applicable to crawler crane support calls for demonstrating the adequacy of the ground and mats by swinging the crane 360° with the boom raised to its greatest angle and with no load on the hook. The rule states that the maximum track bearing pressure that occurs in this condition is greater than the maximum pressure that will occur during any lifting operation.

We know from the crane load studies presented in Chapter 1 that this is not true. An example is shown in Fig. 5.15. Fig. 5.15a shows the track bearing pressures on a soft surface for a Manitowoc 999 Series 3 crane with a 160' (48.8 meters) boom lifting at a radius of 60 feet (18.3 meters). The rated load for this case is 107,000 pounds (48,534 kilograms) and the lift used in this analysis is 80% of rated load, or 85,600 pounds (38,828 kilograms). The maximum track bearing pressure is found to be 51.2 psi (7,373 psf; 353.0 kPa).

Fig. 5.15b shows the track bearing pressures for this crane at its minimum radius for this boom length (28 feet; 8.5 meters) and lifting 4,000 pounds (1,814 kilograms), representative of just a load block and reeving. The maximum track bearing pressure is found to be 45.1 psi (6,494 psf; 311.0 kPa).

Thus, we find that the track bearing pressure that results from a lift of 80% of the rated load in this particular configuration is 113.5% of the "no hook load" track bearing pressure. Other lifts within the load chart will produce even greater track bearing pressures. Further, as is seen in the studies of crawler track bearing pressure patterns shown in Chapter 1, the increase in maximum track bearing pressure that corresponds to an increase in lifted load is nonlinear. Continuing with this example with a Manitowoc 999 Series 3 crane, a 10% increase in load

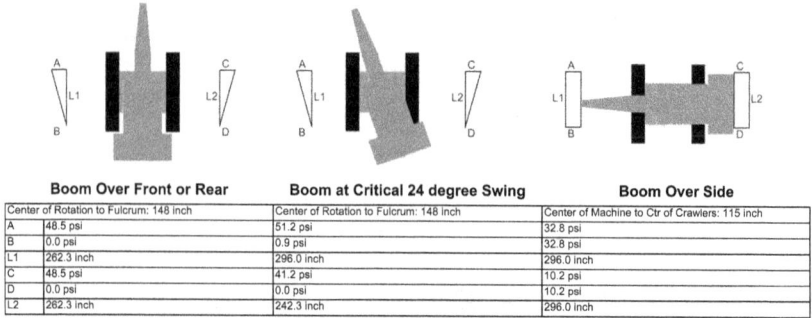

Boom Over Front or Rear		Boom at Critical 24 degree Swing		Boom Over Side	
Center of Rotation to Fulcrum: 148 inch		Center of Rotation to Fulcrum: 148 inch		Center of Machine to Ctr of Crawlers: 115 inch	
A	48.5 psi	A	51.2 psi	A	32.8 psi
B	0.0 psi	B	0.9 psi	B	32.8 psi
L1	262.3 inch	L1	296.0 inch	L1	296.0 inch
C	48.5 psi	C	41.2 psi	C	10.2 psi
D	0.0 psi	D	0.0 psi	D	10.2 psi
L2	262.3 inch	L2	242.3 inch	L2	296.0 inch

(a) Track Bearing Pressures Lifting 80% of Rated Load at a 60' Radius

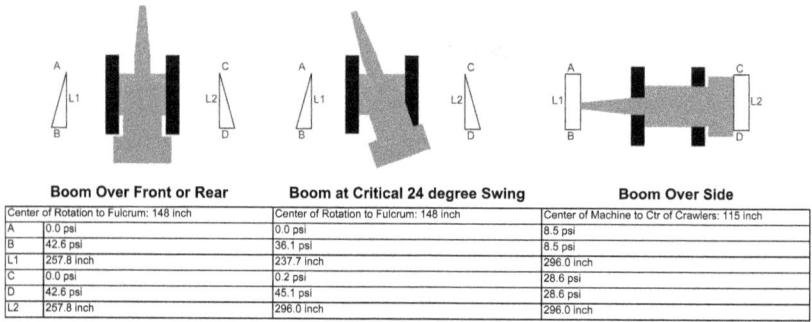

Boom Over Front or Rear		Boom at Critical 24 degree Swing		Boom Over Side	
Center of Rotation to Fulcrum: 148 inch		Center of Rotation to Fulcrum: 148 inch		Center of Machine to Ctr of Crawlers: 115 inch	
A	0.0 psi	A	0.0 psi	A	8.5 psi
B	42.6 psi	B	36.1 psi	B	8.5 psi
L1	257.8 inch	L1	237.7 inch	L1	296.0 inch
C	0.0 psi	C	0.2 psi	C	28.6 psi
D	42.6 psi	D	45.1 psi	D	28.6 psi
L2	257.8 inch	L2	296.0 inch	L2	296.0 inch

(b) Track Bearing Pressures with No Hook Load at Minimum Radius

Figure 5.15 Manitowoc GBPE Output for Critical Track Pressures

(that is, going from 80% of rated load to 88%) results in a maximum track bearing pressure of 58.3 psi, an increase of 13.9% from the bearing pressure at 80% of rated load. Now the crane is at 129.3% of the "no hook load" track bearing pressure. We can see, then, that this oft repeated rule of thumb is not accurate and must not be depended upon for lift planning.

Even if this rule of thumb worked, there is still a serious shortcoming. What if the swing was made and the ground was found to be inadequate? At best, the intended lift plans would have to be revised to account for the lower than expected ground bearing capacity. At worst, the ground would give way and the crane would overturn. Trying to figure out the crane support demands and the ground bearing capacity after the crane is set up and ready to go to work is clearly not an acceptable way to manage a job.

5.6.2   Outrigger-Supported Cranes

A few different rules of thumb are commonly suggested for sizing the mats under outrigger-supported cranes. The most common of these are discussed in this section.

360° Swing. The first rule is a variation of the rule discussed in the previous section for crawler cranes. That is, swing the crane boomed up tight with no load on the hook; the maximum outrigger load during this operation is greater than the maximum outrigger load that will occur during a lift. Again with reference to the support load patterns investigated in Chapter 1, we know that this isn't true. And again, we will provide an example to illustrate this point.

Consider a Link-Belt RTC-80100 Series II crane fitted with full counterweight and a 90-foot (27.4-meter) boom working at a radius of 50 feet (15.2 meters). The rated load in this configuration is 35,700 pounds (16,190 kilograms). Fig. 5.16a is the output from the Link-Belt Ground Bearing Information web site for this lift showing the maximum outrigger load for lifting 80% of this rated load. We see that the maximum outrigger load is 93,146 pounds (42,250 kilograms).

Fig. 5.16b is the web site output for the maximum outrigger load lifting only 2,000 pounds (907 kilograms), representative of the weight of a load block and a small amount of reeving, with the crane boomed up to its minimum radius of 15 feet (4.6 meters). Here, the maximum outrigger load is only 43,637 pounds (19,793 kilograms).

Thus, the outrigger load when lifting 80% of the rated load for the particular configuration defined is 213.5% of the "no hook load" case. Again we see that this rule of thumb is extremely inaccurate and, therefore, unsafe. And as with the crawler crane discussion, trying to sort out the crane support needs after the crane is set up is inefficient as well as potentially unsafe.

Mat Area Based on Rated Load. Another very common rule of thumb relates the required size of outrigger mat to the rated load of the crane. The typical form of this rule states that the required outrigger size in square feet is equal to the maximum rated load of the crane in short tons divided by some number, most commonly 5. In equation form, this rule of thumb is expressed as Eq. 5.1.

$$\text{Mat Area (sq. ft.)} = \frac{\text{Rated Load (short tons)}}{5} \tag{5.1}$$

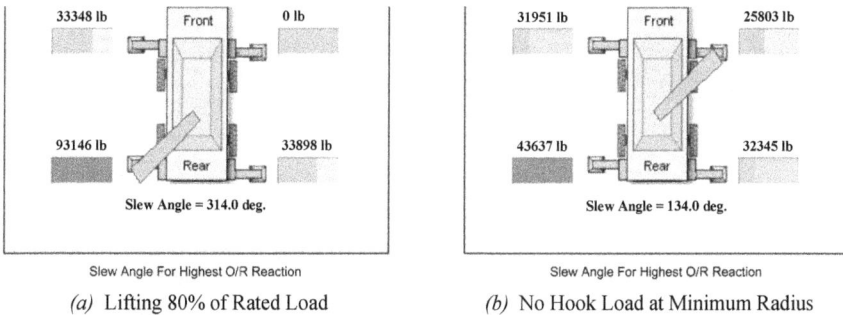

Slew Angle For Highest O/R Reaction

(a) Lifting 80% of Rated Load

(b) No Hook Load at Minimum Radius

Figure 5.16 Link-Belt Ground Bearing Site Output for Maximum Outrigger Loads

In this rule and in those to follow, the term "rated load" refers to the maximum load rating of the crane, not the rated load in the particular configuration and radius to be used for the lift under consideration (e.g. the rated load of the Link-Belt RTC-80100 Series II crane discussed in the previous rule is always 100 tons).

There is one immediate clue that this is not a valid rule. The equation does not account for the strength of the supporting surface (e.g. the allowable ground bearing pressure). Even if we had a simple expression like this that could reliably tell us the maximum outrigger load, we still must know the allowable ground bearing pressure in order to size the mats.

As an example, consider the Link-Belt RTC-80100 Series II crane discussed with respect to the previous rule of thumb. This is a 100-ton crane, so Eq. 5.1 gives us a required outrigger mat area of 100 / 5 = 20 square feet. The outrigger loads output in Fig. 5.15a shows a maximum outrigger load of 93,146 pounds for the particular case discussed. If an outrigger mat that provides 20 square feet of bearing area is used with this crane, the maximum ground bearing pressure will be 93,146 / 20 = 4,657 pounds per square foot. In some areas, this may be acceptable. In others, such as the author's home location near the Texas Gulf Coast, this pressure could exceed the ultimate bearing capacity of the soil.

Variations of Eq. 5.1 seen by the author have the constant in the denominator as low as 4 and as great as 7. The smaller the number, the larger will be the required area given by the equation. Regardless, this rule of thumb is not accurate in that it does not account for the bearing capacity of the ground. Sometimes the result may be adequate, particularly in areas with stronger soils. Other times, the soil will be overloaded.

Three Times the Float Area. This simple rule of thumb states that the required outrigger mat area is three times the area of the outrigger floats with which the crane is equipped by the manufacturer.

Consider three crane models from Grove: the 75-ton (68 tonne) RT875E, the 90-ton (82-tonne) RT890E, and the 130-ton (118-tonne) RT9130E. All three of these cranes are equipped with the same 30.5" (775 mm) diameter round outrigger floats. Clearly, the actual outrigger loads will differ among these three cranes, even when lifting the same load at the same radius. And again, this rule of thumb does not take into account the allowable ground bearing pressure q_a. Thus, we can easily see that this is not an acceptably accurate guide for the sizing of outrigger mats.

Outrigger Load Approximations. There are three somewhat common rules of thumb that incorporate the allowable ground bearing pressure q_a. Looked at another way, these three rules approximate the outrigger load and then divide that value by the allowable ground bearing pressure to arrive at the required outrigger mat area. These rules of thumb are expressed as Eqs. 5.2, 5.3, and 5.4.

$$\text{Mat Area} = \frac{0.65 \times \left(\text{Crane Weight} + \text{Lifted Load}\right)}{q_a} \qquad (5.2)$$

$$\text{Mat Area} = \frac{0.85 \times \text{Rated Load}}{q_a} \qquad (5.3)$$

$$\text{Mat Area (sq. ft.)} = \frac{1,100 \times \text{Rated Load (short tons)}}{q_a \text{ (psf)}} \qquad (5.4)$$

Eq. 5.4 can be written as Eq. 5.5 to make it dimensionally independent.

$$\text{Mat Area} = \frac{0.55 \times \text{Rated Load}}{q_a} \qquad (5.5)$$

In each of these four equations, the numerator is an approximation of the outrigger load. This value is then divided by the allowable ground bearing pressure to arrive at the required outrigger mat area. Eqs. 5.2, 5.3, and 5.5 are dimensionally independent. For example, if the crane's rated load is in pounds and the allowable ground bearing pressure is in pounds per square foot, the calculated required mat area will be in square feet.

The question to be answered now is: Which, if any, of these equations yield a reasonable approximation of the outrigger loads that will be imposed by the crane?

Eq. 5.2 is lift-specific in that it accounts for the weight of the lifted load. We can examine the accuracy of this equation by calculating the maximum outrigger loads for a variety of cranes and lift configurations using the manufacturers' software. This is done for the rated load at the chosen radius and boom length, thereby determining the maximum outrigger load for the configuration. The estimated outrigger load is then calculated using numerator of Eq. 5.2. The two values are then compared. Fig. 5.17 is a plot of 60 such calculations for a range of Grove and Link-Belt cranes with load ratings ranging from 30 tons (27.2 tonnes) to 150 tons (136 tonnes). Outrigger loads were calculated for four different configurations of 15 different cranes. The results are expressed as the outrigger load calculated using the crane manufacturer's software divided by the outrigger load calculated using Eq. 5.2. A value less than 1.00 indicates that the result of Eq. 5.2 is conservative.

As can be seen in Fig. 5.17, Eq. 5.2 generally overestimates the outrigger load, sometimes by as much as a factor of two. The few cases in which the outrigger load from Eq. 5.2 was less than the load calculated with the manufacturer's software, the error was small, not more than about 4%. Thus, the approximation given by Eq. 5.2 will often be wasteful, but it will most likely not be dangerous.

Eqs. 5.3 and 5.5 are variations of the same idea, which is that the maximum outrigger load is equal to a percentage of the maximum rated load of the crane. Eq. 5.4 was derived by the author in 2008 for a client's specific needs, but has since worked its way into the industry and is now applied more generally by some crane

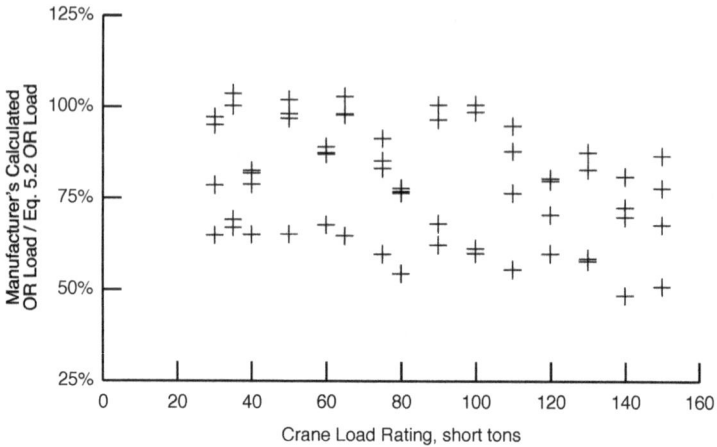

Figure 5.17 Evaluation of Outrigger Loads from Eq. 5.2

users. The basic premise of Eq. 5.5 is that for routine use of a small to medium size mobile crane, the outrigger loads developed likely will not exceed 75% of the maximum outrigger load that the crane will impose under the worst conditions.

Eq. 5.5 is studied here by comparing 75% of the maximum outrigger load from 60 different cranes from Grove, Link-Belt, and Tadano with load ratings ranging from 8.5 tons (7.7 tonnes) to 150 tons (136 tonnes). The results, plotted in Fig. 5.18, are expressed as 75% of the maximum outrigger load given in the crane's specifications divided by the outrigger load calculated using Eq. 5.5. Again, a value less than 1.00 indicates that the result of Eq. 5.5 is conservative.

Fig. 5.18 shows us that Eq. 5.5 is unconservative for the smaller cranes (rated loads less than 40 tons), sometimes significantly so. Eq. 5.5 generally overestimates the outrigger loads for the larger cranes in this study. In additional to the lack of

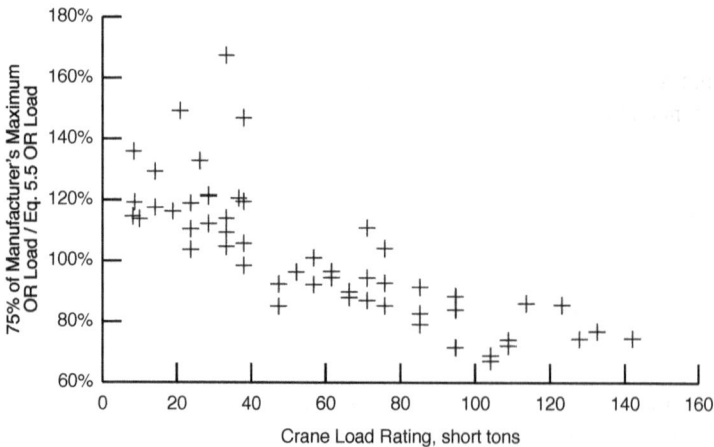

Figure 5.18 Evaluation of Outrigger Loads from Eq. 5.5

accuracy, these results show the problem with taking a method developed for a specific purpose (in this case, as a part of a single company's crane management policy) and applying it generally to all applications.

While some may try to refine Eqs. 5.2, 5.3, and/or 5.5 to improve their accuracy, the experience of the author and others indicates that this is not a reasonable endeavor. The many differences among cranes, even from the same manufacturer, make the development of a simple formula for the estimation of the maximum outrigger load impractical. Add to this the availability of outrigger load calculators from the crane manufacturers and statements in the cranes' specifications that give the maximum outrigger load and we see that the need for a rule of thumb formula simply isn't there. If a quick answer is needed, the safest approach is to look up the maximum outrigger load in the crane's specifications and work with that. Assuming a reduced maximum load for a particular project remains reasonable.

5.7   NOTES ON PRACTICAL APPLICATIONS

By now, the reader must recognize that the design of the supports for a mobile crane is not an exact science. The calculation of the crane's support reactions (outrigger loads or crawler track bearing pressures) are approximate, the strength of soil and crane mats (particularly timber mats) are variable, and the interaction between mats and soil is a very complex behavior that does not provide a constant factor of safety when the design follows practical methods, as outlined in this handbook. Thus, this work must be tempered with a healthy measure of experience-based engineering judgment. This section will provide a few comments on the practical application of the methods discussed in this handbook.

When designing supports for outrigger-supported cranes, the maximum outrigger load typically will not be the same at all locations. This is particularly true if the crane will make one major lift from one position with only a small defined swing. Thus, the required crane mat size may be different for each outrigger. This notwithstanding, it is usually best to specify the same mats under all outriggers. Doing so eliminates the possibility of making an error in mat placement. If this is not possible due to area restrictions or other considerations, then particular care must be taken in laying out the mats to assure that the support demand at each outrigger will be provided.

Chapter 1 discusses the use of a layer of sand or other fine granular material over crane mats to increase the bearing area of a crane's crawler tracks. This technique is not advisable if multiple lifts will be made or if the crane must travel repeatedly. In such a case, the cyclic loading and unloading may cause the sand to be worked out from under the tracks, thus losing its effectiveness. The use of sand for this function is best limited to a short-term crane setup.

Last, the lift planner/engineer must recognize the differences between a single-lift crane setup and a long-term installation. The engineering for a single-lift (or a small group of well defined lifts) crane setup can be performed with a relatively high degree of confidence in the knowledge of the acting loads and the site conditions. A long-term installation in which the crane will make numerous lifts, perhaps not all known at the time the installation is planned, brings with it greater uncertainty in the loads that may be imposed by the crane. Further, the site conditions can change; one heavy rain can materially alter the soil bearing capacity under the mats. Somewhat more conservative assumptions may be appropriate for a long-term installation design.

5.8   INSPECTION OF CRANE MATS

Crane mats must be inspected, just like any other piece of lifting equipment. Unlike any other piece of lifting equipment, however, there are no well defined rejection criteria for mats. What is acceptable and what is cause for rejection are entirely subjective. The responsibility for performing inspections of crane mats and for determining the rejection criteria may fall to the lift director, the crane operator, or the supplier of the mats. This will depend on the party that provides the mats to a particular job and the management structure of that job. Ideally, every owner of crane mats, particularly businesses that stock mats for rental, will implement a program to inspect mats and remove from service those that are deemed unsuitable for continued use.

The strength of a conventional timber crane mat is affected by the condition of the individual timbers. Timbers degrade from exposure to the elements and from infestations of certain types of insects. In all cases, the wood is ultimately attacked by organisms that fall into four categories. These are fungi, insects, bacteria, and marine borers.

Some types of fungus damage is mainly visual (staining) and does not affect the strength of the wood significantly. Other types of fungus damage result in decay of the heartwood, which does result in a loss of strength. Of the species used for mat construction, oak and the tropical hardwoods possess the greatest natural resistance to decay. Insects that pose a threat to the integrity of crane mat timbers include carpenter ants, dry-wood termites, and wood-boring weevils. Damage from bacteria living in the wood is generally not a concern in that the effect on strength properties is slight. Last, marine borers are a concern only to wood structures in salt or brackish waters.

All of the organisms noted here are more active in warmer and more humid climates. Thus, these issues are major concerns of crane mat users in the author's home location of Houston, but not so much so in, for example, Minneapolis. Mat use, storage, and inspection practices must take into account the effects of regional climate differences. More information on the effects of these different organisms on wood, including photographs of typical wood damage, can be found in Chapter 14 "Biodeterioration of Wood" in the *Wood Handbook* (FPL 2010).

One of the major advantages of synthetic mats is that the materials from which they are made generally are impervious to weathering and to insects. That does not mean, however, that synthetic mats are indestructible. Synthetic mats can be damaged by overloading, mishandling, and other such abuses. These types of damage are usually obvious in a visual inspection and typically manifest themselves as cracks or permanent deformation. Since synthetic mats are proprietary products, additional guidance on inspection and removal from service should be obtained from the mat manufacturer.

Last are fabricated steel mats. The inspection of steel mats can be performed just as one would inspect any other steel fabrication. The inspector must look for cracks in both the base metal and any welds, missing bolts or other mechanical fasteners, and local permanent deformations. The types of flaws that would be cause for rejection of, for example, a lifting beam can be applied to the evaluation of a steel crane mat, as well. When flaws are discovered, a qualified person must determine if the mat can be repaired or if it must be permanently removed from service.

Regardless of the type of mat, all crane mats must be periodically inspected and removed from service if determined to be inadequate. In the absence of specific criteria from regulatory agencies or industry standards, responsibility for the development of the frequency of inspections and rejection criteria falls to the crane mat owner/supplier, the lift director, or the crane operator. This is not a responsibility to be taken lightly.

5.9   APPLICATIONS TO OTHER TYPES OF EQUIPMENT

As indicated by the title, the content of this handbook is directed to users of mobile cranes. However, many of the principles developed here can be applied to the design of supports for other types of lifting and load handling equipment, such as telescopic hydraulic gantry systems and skidding systems (also known as jack and slide systems).

Crane mats, as well as individual hardwood timbers, are often used as a part of the support of load handling equipment other than mobile cranes. The analysis methods and allowable stresses developed for mobile crane support design are generally applicable to the design of supports for these other types of equipment. However, direct use without considering the differing demands of other types of equipment may not always be appropriate. For example, telescopic hydraulic gantry systems are very sensitive to out-of-level supports, more so than most types of mobile cranes. Therefore, more stringent limits on allowable ground bearing pressure and mat deflection may be necessary. (Note that this difference may affect the effective length calculation in Eqs. 4.29 through 4.31.)

In summary, many aspects of the methods developed in this handbook for mobile crane support design may be useful in the engineering of supports for other types of lifting and load handling equipment. However, the differences in support requirements presented by these other types of equipment must be accounted for in any such usage.

5.10   APPLICATION SUMMARY – A PERSONAL NOTE

I have been giving lectures on topics related to specialized lifting and rigging since 1991. I have occasionally closed some of my more technical presentations with the following quote from one of my grad school textbooks (Lin and Burns 1981):

> To engineers who, rather than blindly following the
> codes of practice, seek to apply the laws of nature.

This quote is the dedication at the beginning of the book and very nicely sums up the basic philosophy that every engineer and lift planner should follow. The forces that a mobile crane imposes upon its supports are what they are for a given crane configuration and lifted load. Gravity, inertia, wind, and other natural effects all contribute and cannot be ignored. A true understanding may reveal that one or more of these effects is trivial, as has been discussed in Chapter 1, but simply ignoring something doesn't make it go away.

Similarly, the distribution of those support forces to the underlying components, whether they are crane mats, outrigger pads, a concrete slab or any other structural element, must recognize how everything interacts. "This is how we do it" will not alter the reality of structural behavior. When corporate policy and the laws of nature collide, the laws of nature always win.

Every lift planner and lift specialist engineer is strongly encouraged to "seek to apply the laws of nature." Professional obligations also require us to follow the codes of practice, but we need not do so blindly. Truly understanding how cranes, mats, soil, etc. behave under the changing loading from a lifting operation is essential to produce a safe lift.

5.11   REFERENCES

American Association of State Highway and Transportation Officials (AASHTO) (2017), *AASHTO LRFD Bridge Design Specifications*, 8th ed., Washington, D.C.

American Bureau of Shipping (ABS) (2017), *Rules for Building and Classing Steel Barges*, Houston, TX.

American Concrete Institute (ACI) (2014), 318-14 *Building Code Requirements for Structural Concrete and Commentary*, Farmington Hills, MI.

American Institute of Steel Construction (AISC) (2016), *Specification for Structural Steel Buildings*, Chicago, IL.

American Society of Civil Engineers (ASCE) (2015), SEI/ASCE 37-14 *Design Loads on Structures During Construction*, Reston, VA.

American Society of Mechanical Engineers (ASME) (2018), ASME B30.5-2018 *Mobile and Locomotive Cranes*, New York, NY.

Asphalt Institute (1999), MS-1 *Thickness Design – Highways & Streets*, 9th ed., Lexington, KY.

Farny, J.A., and Tarr, S.M. (2008), EB075 *Concrete Floors on Ground*, 4th ed., Portland Cement Association, Skokie, IL.

Forest Products Laboratory (FPL) (2010), General Technical Report FPL-GTR-190 *Wood Handbook – Wood as an Engineering Material*, Madison, WI.

Lin, T.Y., and Burns, N.H. (1981), *Design of Prestressed Concrete Structures*, 3rd ed., John Wiley & Sons, Inc., Hoboken, NJ.

Occupational Safety and Health Administration (OSHA) (2018), Title 29, *Code of Federal Regulations*, "Subpart CC – Cranes and Derricks in Construction," United States Government Printing Office, Washington, D.C.

Shapiro, L.K., and Shapiro, J.P. (2011), *Cranes and Derricks*, 4th ed., The McGraw-Hill Companies, Inc., New York, NY.

Young, W.C., Budynas, R.G., and Sadegh, A.M. (2012), *Roark's Formulas for Stress and Strain*, 8th ed., The McGraw-Hill Companies, Inc. New York, NY.

Appendix 1
Glossary of Specialized Terms

Adjustment Factor   A number by which strength properties are divided to arrive at allowable strength or allowable stress values. This term is used primarily in timber design.

Allowable Ground Bearing Pressure (q_a)   The maximum permissible unit pressure, typically expressed in pounds per square foot (psf) or pascals (Pa), that may be imposed on the supporting surface. This value may be equal to the Ultimate Soil Bearing Pressure divided by a suitable safety factor or it may be a lesser value as limited by other considerations, such as the strength of buried pipes. See Ground Bearing Pressure; see Soil Bearing Capacity; see Ultimate Soil Bearing Pressure.

Angle of Internal Friction (ϕ)   An experimentally determined measure of the ability of soil to resist shear stress.

Angle of Repose   The steepest angle to horizontal to which a material can be piled without slumping.

Bedding   Prepared surface on which the bottom of a buried pipe bears.

Bedding Factor (B_f)   The ratio of the strength of a buried concrete pipe as measured in a three-edge bearing test to the strength of the pipe under the installed condition of loading and bedding.

Blocking   Wood or other material used to support equipment or a component and distribute loads to the ground. It is typically used to support lattice boom sections during assembly/disassembly and under outrigger and stabilizer floats. (This definition is taken from 29 CFR 1926.1401.) Also called dunnage or cribbing.

Bumper Outrigger   See Front Outrigger.

Check   A separation of the wood along the fiber direction that usually extends across the rings of annual growth.

Clear   Free of defects or imperfections. This term is used primarily in timber design.

Common Species   Hardwood and softwood timber species widely used for the construction of crane mats in the United States and Canada. The common species include Douglas fir, hemlock, hickory, beech, and various species of oak. These species are native to North America.

Consolidation   Compression of soil due to applied loading that occurs relatively slowly over a long period of time. See Settlement.

Cover (*H*)   The vertical dimension from the top of a subsurface structure, such as a buried pipe, to the ground surface.

Crane Mat   An assembly of timbers or metal structural members arranged to distribute the loads from a crane's tires, outriggers, or crawler tracks over a large area in order to reduce the ground bearing pressure. Also called pontoon.

Crane Pad   An area of compacted or otherwise specially prepared soil, concrete, wood timbers or mats, or steel plate designed to support a mobile crane.

Cribbing   See Blocking.

Decay   Decomposition of the wood substance caused by the action of wood-destroying fungi, resulting in the softening, loss of strength and weight, and often in change of texture and color. The levels of decay may be described as advanced (or typical) decay (the older stage of decay in which the destruction is readily recognized), incipient decay (the early stage of decay that has not proceeded far enough to impair the hardness of the wood), or pocket rot (the advanced decay that appears in the form of a hole, pocket, or area of soft rot usually surrounded by apparently sound wood).

Design Factor   The ratio of the limit state strength or stress(es) of an element to the permissible internal force(s) or stress(es) created by the external force(s) that acts upon the element. See Adjustment Factor.

Dressed Lumber   See Surfaced Lumber.

Dunnage   See Blocking.

Edge Knot   See Knot.

Effective Bearing Area   The area under a crane mat that is effective in distributing the applied load to the underlying surface, typically equal to the effective bearing length of the mat multiplied by its width.

Effective Bearing Length (L_{eff})   The length of a crane mat as limited by its bending strength, shear strength, or bending stiffness that is assumed in analysis to distribute the applied load to the underlying surface as a uniformly distributed pressure.

Flexible Pavement   Pavement that remains in contact with and distributes loads to the subgrade.   Flexible pavement depends on aggregate interlock, particle friction, and cohesion for stability.   See Rigid Pavement.

Flexible Pipe   A buried pipe that will deflect at least 2% without structural distress.   Corrugated and plain steel pipes are flexible pipes.   See Rigid Pipe.

Float   The rigid structural component that attaches to the outer end of a mobile crane's outrigger to provide some load spreading of the outrigger load to the supporting surface.   The floats are provided by the crane manufacturer.   Also called outrigger float, outrigger pad, outrigger pan, or pad.

Front Outrigger   A hydraulic cylinder or mechanical jack mounted on the front bumper of a truck crane carrier to provide additional stability and to extend the crane's working range over the front of the carrier.   Also called bumper outrigger.

Green   Freshly sawed wood, or wood that has received no drying; unseasoned wood; lumber that may have become wet to above the fiber saturation point may be referred to as being in the "green condition;" also, wood above a stipulated moisture content, such as lumber above a moisture content of 19%.

Ground Bearing Pressure (q)   The unit pressure, typically expressed in pounds per square foot (psf) or pascals (Pa), that a crane imposes on the supporting surface. See Allowable Ground Bearing Pressure.

Ground Conditions   The ability of the ground to support the equipment (including slope, compaction, and firmness).   (This definition is taken from 29 CFR 1926.1402.)

Hardwood   One of the botanical groups of trees that have broad leaves in contrast to the conifers or softwoods.   The term "hardwood" has no reference to the actual hardness of the wood.   See Softwood.

Heart Shake   See Shake.

House   See Superstructure.

Knot   A portion of a branch or limb that has been surrounded by subsequent growth of the wood of the tree. An *edge knot* is a knot that is located at the edge of the face of a piece of lumber.

Lateral Earth Pressure   The horizontal pressure from a mass of soil that acts on a vertical surface as a result of vertical loads (both the soil's self weight and any load supported on the ground surface).

Latewood   The denser, smaller celled, later-formed part of a growth layer Also called summerwood.

Limit State   A condition in which a structure or component becomes unfit for service and is judged either to be no longer useful for its intended function (called a *serviceability limit state*) or to have reached its ultimate load-carrying capacity (called a *strength limit state*). (This definition is based on the AISC *Specification for Structural Steel Buildings*.) Limit states that affect the design of mobile crane supports include crane mat bending strength, crane mat shear strength, crane mat deflection, soil strength, and soil deformation.

Machinery House   See Superstructure.

Modulus of Soil Reaction (E')   The elastic lateral stiffness of soil with respect to the support provided by the soil to a flexible buried pipe; expressed in units of force divided by area [i.e., pounds per square inch (psi) or newtons per square millimeter (N/mm^2); one N/mm^2 equals one megapascal (MPa)].

Modulus of Subgrade Reaction (k_s)   The elastic stiffness of soil computed as pressure divided by deflection; expressed in units of force divided by volume [i.e., pounds per cubic foot (pcf) or newtons per cubic meter (N/m^3)].

Moisture Content   The amount of water contained in a piece of wood, usually expressed as a percentage of the mass of the ovendry wood.

Outrigger Float   See Float.

Outrigger Pad   A wood, metal, or synthetic structural element that is placed on the supporting surface and on which bears the crane's float that is used to distribute the outrigger load over a larger area. See Float.

Outrigger Pan   See Float.

Pad   See Crane Pad; see Float; see Outrigger Pad.

Pin Hole   A hole not over 1/16" (1.6 mm) in diameter in a piece of lumber.

Pitch Streak   A well-defined accumulation of pitch in a more or less regular streak in the wood.

Pith   The small, soft tissue occurring in the structural center of a tree trunk, branch, twig, or log.

Pocket   A well-defined opening between the rings of annual growth which develops during the growth of the tree. A pocket usually contains pitch or bark.

Pocket Penetrometer   A handheld tool used for making field classifications of cohesive soils in terms of consistency, shear strength and approximate unconfined compressive strength.

Pontoon   See Crane Mat; see Float.

Posts and Timbers   Lumber of square or approximately square cross section, 5" x 5" (125 mm x 125 mm) (nominal) or larger, with its width not more than 2" (50 mm) greater than its thickness. Although graded for use as columns, lumber of this type is commonly used for crane mat construction.

Presumptive Bearing Pressure   An allowable ground bearing pressure based on soil classification that is specified by a building code or design standard.

Proctor Test   Standardized test methods for measuring the maximum dry unit weight and optimum moisture content for soil. The methods are defined by ASTM D698 for the standard Proctor test and D1557 for the modified Proctor test.

Rigid Pavement   Pavement that provides high bending strength and stiffness, allowing it to distribute loads to the subgrade over a comparatively large area. See Flexible Pavement.

Rigid Pipe   A buried pipe with a stiffness greater than that to be treated as a flexible pipe. Concrete, clay, and ductile iron pipes are rigid pipes. See Flexible Pipe.

Ring Shake   See Shake.

Rough Lumber   Lumber that has not been dressed (surfaced) but has been sawed, edged, and trimmed. The actual dimensions of rough lumber are approximately equal to the nominal dimensions. Also called rough sawn. See Surfaced Lumber.

Seasoned   Wood that has been dried, either by exposure to the air or by heating in a kiln, to remove moisture. Seasoned wood generally has a moisture content of 19% or less, depending on the species and intended use

Settlement   Compression of soil that occurs immediately upon application of load. See Consolidation.

Shake   A longitudinal separation of the wood. *Heart shake* is shake that starts out at or near the pith and extends radially. *Ring shake* is shake that occurs in the plane of the growth rings in the outer portion of the latewood for partial or entire encirclement of the pith, occasionally moving radially to an adjacent latewood ring.

Skip   An unsurfaced area on dressed lumber. Also called skip dressing.

Slope of Grain   The slope of the grain of a piece of lumber to the sides of the piece.

Softwood   One of the botanical groups of trees that in most cases have needlelike or scalelike leaves; the conifers; also the wood produced by such trees. The term "softwood" has no reference to the actual hardness of the wood. See Hardwood.

Soil Bearing Capacity (q_{ult})   Bearing capacity of a soil mass as limited by the strength and stiffness of the soil. Also called Ultimate Soil Bearing Pressure. See Allowable Ground Bearing Capacity.

Split   A separation of the wood parallel to the fiber direction, due to the tearing apart of the wood cells.

Stain   A discoloration in wood that may be caused by microorganisms, metal, chemicals, or other agencies.

Standard Penetration Test   A method of measuring the strength of soil by pushing a cone penetrometer into the soil. Specific details of the design of the penetrometer and the test methods are established by industry (e.g. ASTM) standards.

Strength Ratio   The hypothetical ratio of the strength of a piece of lumber to the strength it would have if no weakening defects were present. Strength ratio values are always less than 1.00.

Stress Ratio   The ratio of the actual stress in an element due to the acting loads divided by the corresponding allowable stress. See Utilization Ratio.

Summerwood   See Latewood.

Superstructure   The rotating part of a mobile crane that contains the operating machinery and on which is mounted the operator's cab. Also called the house, the machinery house, the upper, or the upperworks.

Supporting Materials   Blocking, mats, cribbing, marsh buggies (in marshes/ wetlands), or similar supporting materials or devices. (This definition is taken from 29 CFR 1926.1402.)

Surfaced Lumber   Lumber that is dressed by running it through a planer. The actual dimensions of surfaced lumber from mills in the U.S. are commonly 1/4" (6 mm) to 3/4" (19 mm) less than the nominal dimensions. Lumber surfaced on all four faces is designated S4S (surfaced four sides). Also called dressed lumber or planed lumber. See Rough Lumber.

Torn Grain   An irregularity in the surface of a piece of wood where wood has been torn or broken out by surfacing. Heavy torn grain may be up to 1/8" (3.2 mm) deep.

Track Bearing Pressure   The pressure, typically expressed in pounds per square foot (psf), pounds per square inch (psi), or pascals (Pa), that the track of a crawler crane imposes on its supporting surface. This pressure may be uniform along the length of the track or it may very from one end to the other.

Tropical Hardwoods   Hardwood species typically native to Africa and South America used in crane mat construction. The most common tropical hardwoods used for mats include azobe (ekki), eucalyptus, mora, and wamara.

Ultimate Soil Bearing Pressure (q_{ult})   Bearing capacity of a soil mass as limited by the strength and stiffness of the soil. Also called Soil Bearing Capacity. See Allowable Ground Bearing Capacity.

Unsound Wood   A disintegration of the wood substance due to the action of wood-destroying fungi.

Upper   See Superstructure.

Upperworks   See Superstructure.

Utilization Ratio   The ratio of the actual force or stress in an element due to the acting loads divided by the corresponding allowable force or stress. Also called a Stress Ratio when the values used are stresses.

Wane   Bark or lack of wood from any cause on an edge or corner of a piece of lumber.

White Speck   Small white or brown spots in the wood caused by a fungus. White speck develops in the living tree and does not develop further in wood in service.

Appendix 2
USCU / SI Conversion Factors

CONVERSION FACTORS

Following are conversion relationships for the most commonly encountered U.S. customary units (USCU) and International System of Units (SI) quantities applicable to the use of mobile cranes. The standard abbreviations for the SI quantities are shown in parentheses. These factors are as defined in Annex A of IEEE/ASTM SI 10-2016 *American National Standard for Metric Practice*. A multiplier in bold type indicates that the conversion factor is exact and, therefore, all subsequent digits are zero. All other conversion factors are expressed with an accuracy of seven significant digits.

To convert from	To	Multiply by
Length		
inches	millimeters (mm)	**25.4**
feet	meters (m)	**0.304 8**
yards	meters (m)	**0.914 4**
miles (U.S. statute)	kilometers (km)	1.609 347
Area		
square inches	square millimeters (mm^2)	**645.16**
square feet	square meters (m^2)	**0.092 903 04**
Volume		
cubic inches	cubic millimeters (mm^3)	**1.638 706 4 E04**
cubic feet	cubic meters (m^3)	0.028 316 85
quarts (U.S. liquid)	liters (L)	0.946 352 9
gallons (U.S.) (231 in^3)	liters (L)	3.785 412
Mass		
pounds	kilograms (kg)	**0.453 592 37**
short tons (2,000 pounds)	metric tons[a] (t) (1,000 kg)	0.907 184 7

To convert from	To	Multiply by
Force		
pounds	newtons (N)	4.448 222
short tons (2,000 pounds)	kilonewtons (kN)	8.896 444
kilograms	newtons (N)[b]	**9.806 65**
Bending Moment		
pound-inch	newton-meter (N-m)	0.112 984 8
pound-foot	newton-meter (N-m)	1.355 818
Pressure or Stress		
pounds per square inch	kilopascal (kPa)	6.894 757
pounds per square inch	megapascal (MPa)[c]	0.006 894 757
pounds per square foot	pascal (Pa)	47.880 26
pounds per square foot	kilograms per square meter (kg/m^2)	4.882 428
Density		
pounds per cubic foot	kilograms per cubic meter (kg/m^3)	16.018 46
pounds per cubic inch	kilonewtons per cubic meter (kN/m^3)[c]	271.447 2
kips per cubic foot	kilonewtons per cubic meter (kN/m^3)[c]	157.087 5
Power		
horsepower (550 ft-lb/s)	kilowatts (kW)	0.745 699 9
foot-pounds/second	watts (W)	1.355 818
Temperature		
degrees Fahrenheit (°F)	degrees Celsius (°C)	t°C = (t°F-**32**) / **1.8**
Unit Weight		
pounds per foot	newtons per meter (N/m)	14.593 90
pounds per foot	kilograms per meter (kg/m)	1.488 164

[a] The metric ton, or tonne, is not a formal SI unit (see SI 10-2016, Table 6), but is in common usage in heavy lifting and rigging work.

[b] This conversion is based on the acceleration due to gravity, equal to 9.80665 m/s^2.

[c] This conversion does not appear in SI 10-2016; it is included here for convenience.

SI DERIVED UNITS

Some engineering calculations express quantities in terms of basic units, rather than derived units. For example, stress may be expressed as pascals or as newtons per square meter. The following table shows the equivalences between commonly encountered expressions. These equivalences are definitions of the indicated quantities and, therefore, are exact.

Quantity	Derived Unit	Expressed in Terms of Other SI Units
Pressure or Stress	megapascal (MPa)	$1 \text{ MPa} = 10^6 \text{ N/m}^2$
Pressure or Stress	megapascal (MPa)	$1 \text{ MPa} = 1 \text{ N/mm}^2$
Pressure or Stress	pascal (Pa)	$1 \text{ Pa} = 1 \text{ N/m}^2$
Pressure or Stress	tonnes per square meter (t/m^2)	$1 \text{ t/m}^2 = 9.806\ 65 \text{ kPa}$

REFERENCE

ASTM International (2016), SI 10-2016 *American National Standard for Metric Practice*, West Conshohocken, PA.

Appendix 3
Rounding Conventions

Many of the tables of timber properties in this handbook were developed through the calculation of values or through the conversion of values originally expressed in U.S. customary units (USCU) to values in SI units. Conversion factors are listed in Appendix 2. The rounding conventions used in the development of these values are defined in this appendix.

USCU TO SI CONVERSIONS

The following discussion is based on the provisions of ASTM (2016), Annex B. This annex provides a set of rules expressed as guidelines and examples, rather than hard numbers, for the rounding of converted values. The fundamental goal with respect to the conversion of the values of interest here (stress, force, weight or mass, and length) is that the converted numbers do not have or imply an accuracy that is greater than that of the source values. The practice to be followed calls for exact calculation of the conversion (i.e., the source value is multiplied by the conversion factor without any rounding) and then the result is rounded to obtain a comparable level of accuracy (i.e., a comparable number of significant digits). Conversion factors from ASTM (2016) that are applicable to the subject matter of this handbook are listed in Appendix 2.

Rounding to an appropriate level of accuracy requires knowledge of the precision of the original value before conversion. For example, consider a measurement of 12 inches. The exact conversion to SI units is 12 inches x 25.4 mm per inch = 304.8 mm. If the measurement of 12 inches is only accurate to one inch, then the appropriate rounded value is 300 mm. If the original measurement is accurate to 1/16 inch, then the appropriate rounded value is 305 mm (rounded to the nearest millimeter).

The original work in this handbook generally was performed using values expressed in USCU, with SI values converted from the USCU values (in a few cases, values were provided in both USCU and SI units in the source material). Wherever the source value has not been rounded or is considered to be exact, no rounding has been employed in the conversion to SI units. Where the source value has been rounded, the rounding convention used in the units conversion is as given on the next page, unless otherwise stated in the text or in a table note. The rounding convention may deviate from the ASTM (2016) guidelines if necessary to maintain values that have been in use in practice, such as material properties shown in recognized standards, and are considered acceptable as is. For example, the conversions of USCU to SI values of clear wood properties shown in Table 3.7 follow the rounding convention used in FPL (2010), in which both USCU and SI values are given for some of the properties listed in this table.

Measured value made to the nearest 1/16"
    Round to nearest millimeter (1/16" = 1.6 mm)

Measured value made to the nearest 1/8"
    Round to nearest 3 millimeters (1/8" = 3.2 mm)

Nominal dimensions given to the nearest inch
    Round to nearest 25 mm (1" = 25.4 mm)

Nominal dimensions given to the nearest foot
    Round to nearest 300 mm for short dimensions (1' = 304.8 mm)

Specified clear wood modulus of rupture
    Round to nearest 100 kPa

Specified clear wood shear strength
    Round to nearest 10 kPa

Specified clear wood compressive stress perpendicular to the grain
    Round to nearest 10 kPa

Specified clear wood side hardness
    Round to nearest 50 N

Specified clear wood modulus of elasticity
    Round to nearest 1,000 kPa

Allowable bending stress
    Round to nearest 50 kPa

Allowable shear stress
    Round to nearest 10 kPa

Allowable compressive stress perpendicular to the grain
    Round to nearest 50 kPa

Modulus of elasticity for design
    Round to nearest 100,000 kPa

TIMBER PROPERTY CALCULATIONS

The following rounding conventions are taken from ASTM (2011), paragraph 6.1.1, for the calculation of allowable properties for timber design from clear wood properties.

Bending Stress
Round to the nearest 50 psi (340 kPa) for allowable stress of 1,000 psi (6.9 MPa) or greater; otherwise round to the nearest 25 psi (170 kPa)

Horizontal Shear Stress
Round to the nearest 5 psi (34 kPa)

Compression Stress Perpendicular to the Grain
Round to the nearest 5 psi (34 kPa)

Modulus of Elasticity
Round to nearest 100,000 psi (690 MPa)

ASTM (2011), paragraph 6.1.1, further states that the rounding rules of ASTM (2016) shall be followed.

REFERENCES

ASTM International (2011), D245-06(2011) *Standard Practice for Establishing Structural Grades and Related Allowable Properties for Visually Graded Lumber*, West Conshohocken, PA.

ASTM International (2016), SI 10-2016 *American National Standard for Metric Practice*, West Conshohocken, PA.

Forest Products Laboratory (FPL) (2010), General Technical Report FPL-GTR-190 *Wood Handbook – Wood as an Engineering Material*, Madison, WI.

Appendix 4
Notation

The symbols used in the equations presented generally have been selected to be consistent with previously published work, including the cited references. In a number of cases, this has resulted in a symbol being used with more than one meaning. It is incumbent upon the responsible engineer to assure that all symbols are used correctly in any calculations performed in the design of mobile crane supports. All symbols used in this handbook are listed here for reference.

A = area (general usage)

A = length of the effective bearing area on a plane at the top of a buried pipe

A = pipe wall area per unit length

A_{reqd} = required crane mat bearing area

A_v = shear area

B = width of the effective bearing area on a plane at the top of a buried pipe

B = crane mat width

B_f = soil load bedding factor for concrete pipe analysis

B_{fLL} = live load bedding factor for concrete pipe analysis

C = actual width of crawler track bearing (w_h or w_s, as appropriate); bearing width of the track or outrigger float

C = compressive thrust in the walls of a buried corrugated pipe

C_D = load duration factor

CL_l = vertical load above a pipe due to the crane load at the surface, expressed as a load per unit length of the pipe

CL_p = vertical pressure above a pipe due to the crane load at the surface

D = outside diameter of a pipe or outside horizontal dimension of arch or elliptical pipe

D = cumulative loading duration in days used in the calculation of C_D

$D_{0.01}$ = D-load of a concrete pipe

D_1 = deflection lag factor for pipe deflection calculations

D_i = inside diameter of a circular pipe or inside span of an elliptical or arch pipe

DL_l = soil dead load above a pipe, expressed as a load per unit length of the pipe

DL_p	=	soil dead load above a pipe, expressed as a pressure
E	=	modulus of elasticity of a structural material (e.g. steel or wood)
E'	=	modulus of soil reaction
E_c	=	clear wood modulus of elasticity
E_s	=	modulus of elasticity of soil
E_x	=	modulus of elasticity of mat layer x (where two or more layers of mats are stacked)
$(EI)_{eq}$	=	equivalent pipe wall stiffness per inch (mm) of pipe length
F	=	size factor for timber design
F_b	=	allowable bending stress
F_c	=	allowable compressive stress in the wall of a corrugated pipe
$F_{c\perp}$	=	allowable compressive stress perpendicular to the grain
F_v	=	allowable shear stress
F_y	=	yield stress
FS	=	factor of safety
H	=	depth of cover over a buried pipe or other subsurface element
H_c	=	clear wood side hardness
I	=	moment of inertia
I_x	=	moment of inertia; major axis moment of inertia
I_x	=	moment of inertia of mat layer x (where two or more layers of mats are stacked)
I_y	=	minor axis moment of inertia
K	=	bedding constant for steel pipe analysis
K	=	factor for corrugated pipe analysis based on depth of cover and soil density
K	=	adjustment factor (structural design factor) for timber design
K_w	=	coefficient for steel plate design
L	=	actual length of crawler track bearing due to applied load, either L_{bh} or L_{bl} as applicable
L	=	length of bearing pressure plane at the top of a buried concrete pipe
L_a	=	shorter distance from the center of the crane loading (outrigger float or crawler track) to the end of an eccentrically loaded mat
L_b	=	longitudinal center of gravity of the boom
L_b	=	longer distance from the center of the crane loading (outrigger float or crawler track) to the end of an eccentrically loaded mat
L_{bh}	=	length of bearing of the more heavily loaded track
L_{bl}	=	length of bearing of the more lightly loaded track
L_{bt}	=	horizontal distance from the center of rotation to the boom tip
L_c	=	longitudinal center of gravity of the counterweight
L_c	=	cantilevered length of the crane mat
L_e	=	effective supporting length of a buried concrete pipe
L_{eff}	=	effective bearing length of the crane mat
L_h	=	longitudinal center of gravity of the boom hoist components
L_j	=	longitudinal center of gravity of the jib

L_{jt} = horizontal distance from the center of rotation to the jib tip

L_l = longitudinal center of gravity of the crane's lower

L_m = longitudinal center of gravity of the jib mast and backstays

L_{mat} = actual length of the crane mat

L_{max} = crane load carried by the most heavily loaded mat layer in an arrangement of two or more layers in which the mat timbers are parallel to one another

L_{mb} = main boom length

L_{reqd} = required effective bearing length of the crane mat

L_s = longitudinal center of gravity of the complete rotating assembly of the crane

L_t = total crane load (where two or more layers of mats are stacked)

L_u = longitudinal center of gravity of the machinery house

L_x = crane load to mat layer x (where two or more layers of mats are stacked)

LCG = longitudinal center of gravity of the complete crane with its lifted load

M = bending moment

M_n = allowable moment

M_{nx} = allowable moment of mat layer x in an arrangement of two or more layers

MR_c = clear wood modulus of rupture

P = vertical reaction between the crane and the supporting surface; may be an outrigger load, a crawler track load, or a portion thereof (e.g. the crane load applied to one mat), depending on usage

P = total pressure at the top of a buried pipe

P_c = allowable total pressure at the top of a buried steel pipe, based on the limit state of ring buckling

P_c = load acting on the cantilevered length L_c of a crane mat

P_l = left crawler track load

P_{LF} = left front outrigger load

P_{LR} = left rear outrigger load

P_p = unit pressure at the plane at the top of the pipe

P_r = right crawler track load

P_{RF} = right front outrigger load

P_{RR} = right rear outrigger load

P_s = concentrated load at the surface

R = pipe outside radius

R_c = rate of change of the track bearing pressure

R_o = outside diameter of round pipe; or outside vertical dimension of arch or elliptical pipe

R_T = ratio of the higher track load to the total crane load

R_{Vc} = rate of change of the track bearing pressure, based on the total crane weight and longitudinal center of gravity

R_w = water buoyancy factor

S	=	5% exclusion bending strength
S_b	=	ratio of actual S_x to nominal S_x
S_v	=	ratio of actual cross-sectional area to nominal cross-sectional area
S_x	=	section modulus; major axis section modulus
S_y	=	minor axis section modulus
T	=	track load, either T_h or T_l as applicable
T_h	=	greater track load; i.e., the greater of P_l or P_r
T_l	=	lesser track load; i.e., the lesser of P_l or P_r
TCG	=	transverse center of gravity of the complete crane with its lifted load
T.E.B.	=	three-edge-bearing strength of a concrete pipe
V	=	sum of the weight of the crane and its lifted loads
V	=	shear
V_c	=	clear wood shear strength
V_n	=	allowable shear strength
V_{nx}	=	allowable shear strength of mat layer x in an arrangement of two or more layers
W	=	self-weight of the crane mat
W_b	=	weight of the boom
W_{bt}	=	total load at the boom tip
W_c	=	weight of the counterweight
W_F	=	weight of fluid in the pipe
W_h	=	weight of the boom hoist components
W_j	=	weight of the jib
W_{jt}	=	total load at the jib tip
W_L	=	crane load on a concrete pipe based on the effective supporting length
W_l	=	weight of the crane's lower
W_m	=	weight of the jib mast and backstays
W_s	=	weight of the complete rotating assembly of the crane
W_u	=	weight of the machinery house
X_{max}	=	soil displacement at which the behavior is assumed to change from elastic to perfectly plastic (used for the calculation of an approximate value of k_s)
a	=	length of bearing of a crane mat on the ground surface above a pipe
a	=	quadratic coefficient in a quadratic equation
b	=	width of bearing of a crane mat on the ground surface above a pipe
b	=	footing dimensions at the surface (for Boussinesq curves)
b	=	linear coefficient in a quadratic equation
c	=	constant in a quadratic equation
c	=	distance from the last track roller to the centroid of the load acting on the cantilevered length of a crane mat that is parallel to the length of the crawler track
c'	=	longitudinal center of gravity from the heavily loaded end of the tracks (this value is always less than or equal to 0.50 d_l)

c_h' = longitudinal center of gravity from the heavily loaded end of the more heavily loaded track

c_l' = longitudinal center of gravity from the heavily loaded end of the more lightly loaded track

d = horizontal distance from load P_s to the point at which P_p is to be determined using the Boussinesq equation

d = depth (or thickness) of a timber or a crane mat

d_l = longitudinal spacing between the front and rear outrigger pairs

d_l = either d_{lh} or d_{ls} as applicable to a particular calculation

d_{lb} = length of crawler track bearing on a hard surface with the sprocket at one end only blocked

d_{lf} = longitudinal distance from the center of rotation to the left front outrigger

d_{lh} = length of crawler track bearing on a hard surface

d_{lr} = longitudinal distance from the center of rotation to the left rear outrigger

d_{ls} = length of crawler track bearing on a soft surface

d_{rf} = longitudinal distance from the center of rotation to the right front outrigger

d_{rr} = longitudinal distance from the center of rotation to the right rear outrigger

d_t = transverse spacing of outriggers or crawler tracks, center-to-center

e = eccentricity of the centroid of the crawler track load to the center of the crawler track bearing length

e_h = eccentricity of the centroid of the load of the more heavily loaded crawler track to the center of the crawler track bearing length

e_m = eccentricity of the centroid of the crawler track load to the center of the bearing length of a longitudinally oriented mat under the track

e_t = eccentricity of the centroid of the crawler track load to the center of the crawler track bearing length

f_b = calculated bending stress

f_c = calculated compressive stress in the wall of a corrugated pipe

f_v = calculated shear stress

h_w = height of water surface above top of pipe

k_s = modulus of subgrade reaction

m_{max} = maximum bearing pressure acting from the underside of a crane mat to the surface below of a mat that is oriented parallel to the length of a crawler track

m_{min} = minimum bearing pressure acting from the underside of a crane mat to the surface below of a mat that is oriented parallel to the length of a crawler track

m_{roller} = bearing pressure acting from the underside of a crane mat to the surface below of a mat that is oriented parallel to the length of a crawler track at a point in line with the last track roller

m_{shear}	=	bearing pressure acting from the underside of a crane mat to the surface below of a mat that is oriented parallel to the length of a crawler track at a point distant from the last track roller by a length equal to the thickness of the mat (timber mat design only)
n	=	total number of mat layers (where two or more layers of mats are stacked)
n_w	=	number of webs (steel or aluminum beam design)
p	=	crawler track bearing pressure; either p_{max} or p_{min}
p	=	ground surface pressure due to the crane
$p_{blocked}$	=	track bearing pressure for a blocked track
p_{max}	=	maximum crawler track bearing pressure
p_{min}	=	minimum crawler track bearing pressure
p_{soft}	=	track bearing pressure calculated based on a soft surface
$p_{V\,max}$	=	maximum track bearing pressure, calculated from the total crane weight and longitudinal center of gravity
$p_{V\,min}$	=	minimum track bearing pressure, calculated from the total crane weight and longitudinal center of gravity
pct	=	fraction (or percentage) of the total crane load carried by the most heavily loaded mat layer in an arrangement of two or more layers in which the mat timbers are parallel to one another
q	=	vertical pressure within the soil mass due to an applied pressure at the surface (for Boussinesq curves)
q	=	pressure applied at the surface for the determination of the modulus of subgrade reaction
q	=	ground bearing pressure due to the crane load P
q_a	=	allowable ground bearing pressure
q_a	=	bearing pressure applied at the surface (for Boussinesq curves)
q_t	=	calculated total ground bearing pressure
q_{ult}	=	ultimate soil bearing pressure
r	=	radius of gyration of the pipe wall
t	=	horizontal distance from the CL of rotation to the boom foot pin
t	=	thickness
t_{max}	=	maximum bearing pressure applied by a crawler track to a crane mat oriented parallel to the length of the track
t_{min}	=	minimum bearing pressure applied by a crawler track to a crane mat oriented parallel to the length of the track
t_w	=	web thickness
w_h	=	width of crawler track bearing on a hard surface
w_s	=	width of crawler track bearing on a soft surface
x_b	=	boom CG location measured along the length of the boom
x_j	=	jib CG location measured along the length of the jib
x_o	=	longitudinal offset of the centroid of the crane's support footprint (crawler bearing area or outrigger group) to the center of rotation
α	=	angle of swing of the crane's superstructure

α	=	coefficient for steel plate design
β	=	coefficient for steel plate design
Δ	=	deflection of a crane mat at the end of L_c
Δy	=	vertical deflection of a buried pipe due to imposed pressure
γ	=	soil density
δ	=	vertical compressive displacement of soil
θ	=	angle of main boom from horizontal
λ	=	ratio of c' to the maximum bearing length of the track
λ_{tp}	=	the limit of λ that defines the point at which both bearing pressure envelopes are trapezoidal
λ_{tr}	=	the limit of λ that defines the point at which both bearing pressure envelopes are triangular
λL_{eff}	=	value that defines the relative stiffness of a beam on soil
μ	=	jib offset angle
μ	=	Poisson's ratio
μ	=	the average (mean) value of the strength property
σ	=	the standard deviation of the strength property
σ_{bw}	=	through-wall bending stress in a steel pipe
ϕ	=	angle of internal friction of soil

INDEX

Symbols

5% exclusion limit 116–117
29 CFR 1926.1401 182
29 CFR 1926.1402 181–182

A

A1A Software, LLC 3D Lift Plan web site 31
AASHTO *LRFD Bridge Design Specifications* 82, 204
ABS. *See* American Bureau of Shipping
ACPA *Concrete Pipe Design Manual* 65–70
Addressing low soil bearing capacity 54–56
Adjustment factor 117
AISI *Modern Sewer Design* 70–73
ALA *Guidelines for the Design of Buried Steel Pipe* 60–62
Allowable Bearing Stress for APA Wood Structural Panels 125
Allowable bending stress (steel pipe) 62
Allowable ground bearing pressure 130, 133, 137, 186
Allowable soil bearing pressure 186
Allowable stresses
 adjustment factor 117
 duration of load 103, 107–110, 118
 moisture content 103
 National Design Specification for Wood Construction 106–115
 size factor 103–104
 steel 123
 strength ratio 96, 116–117
 suggested values for timber crane mats 115, 189
 synthetic materials 124
 timber – based on NDS 106–115
 timber – based on surveys 115–117
ALSC. *See* American Lumber Standard Committee, Inc.
Aluminum Association

Specification for Aluminum Structures 124, 137
Aluminum mats 124
American Association of State Highway and Transportation Officials
 LRFD Bridge Design Specifications 82, 204
American Bureau of Shipping
 Rules for Building and Classing Steel Barges 82, 205
American Concrete Institute
 Building Code Requirements for Structural Concrete 77, 82, 204
American Institute of Steel Construction 123, 155, 172
 Specification for Structural Steel Buildings 77, 82, 123, 137, 155, 172, 204
 Specification for Structural Steel Buildings – Allowable Stress Design and Plastic Design 123, 137, 155, 172
American Lumber Standard Committee, Inc. (ALSC) 95
American Society of Mechanical Engineers B30.5 182–183
Angle of internal friction 52–53
APA – The Engineered Wood Association 125
Applications to other types of equipment 215
Asphalt Institute 82, 205
 MS-1 *Thickness Design – Highways & Streets* 82, 205
ASTM standards 66, 94–95
 C14-15a 66
 C76-16 66
 C506-16b 66
 C507-16 66
 C655-16 66
 C857-16 73
 C858-10e1 73
 C891-11 73
 C985-15 66
 D9-12 94

D245-06(2011) 94–95, 96–104, 107, 116–117, 119
D1196-12(2016) 49
D2555-17 94–95, 103, 111, 116–117, 120
AWWA M11 *Steel Pipe – A Guide for Design and Installation* 60
Azobe 92–93, 95, 111–112

B

B30.5 182–183
Balanced mat analysis method 137–154
  comments on support deflection 150
  design factor provided 146–150
  eccentrically loaded mats 151–154
  effective bearing area calculation method 137–142
  effective bearing length 137–142
Barge-mounted cranes 201–204
Beam on an elastic foundation 129, 143–148
Boussinesq equation 59–60
Boussinesq "pressure bulbs" 43–44
British Standards Institution
  Structural Timber – Visual Strength Grading of Tropical Hardwood 102
Building Code Requirements for Structural Concrete 77, 82, 204
Bulkheads. *See* Retaining walls and bulkheads
Buried pipes 57–73. *See also* Strength of buried pipes
  prism load 57

C

C14-15a 66
C76-16 66
C478-15a 73
C506-16b 66
C507-16 66
C655-16 66
C857-16 73
C858-10e1 73
C891-11 73
C985-15 66
Checks 96–97, 98
Clear wood properties. *See* Clear wood stresses
Clear wood stresses 111, 112, 116–117
  5% exclusion limit 116–117
  common species 111
  tropical species 112
Coefficient of subgrade reaction. *See* Modulus of subgrade reaction

Comments on past mat design methods 135–137
  utilization ratio 135–136
Common species (def.) 106
Concrete Floors on Ground 82, 205
Concrete pipe 65–70
  D-load 68–70
  three-edge-bearing load 70
Concrete Pipe Design Manual 65–70
Consolidation 46, 186
Corrugated steel pipes 70–73
  truncated pyramid 70
Cottonwood 121
Crane loads 1–40
  application summary 39–40
  lifted load 5–6
  reactions from the lower 4–5
  reactions from the superstructure 1–4
  variable load effects 33–39
Crane loads acting on subsurface structures 57–60
  Boussinesq equation 59–60
  buried pipes 57–60
  retaining walls and bulkheads 75–78
  truncated pyramid 58–59, 67–68
Crane mat behavior 129–180
  application summary 177–179
Crane mat design process
  bearing on soil 189–201
  bearing on structures or pavement 201–205
  crane mats on slabs or pavement 201–202
  crane mats under crawler tracks 191–201
  crane mats under outriggers 190
  eccentrically loaded mats 151–154
  notes on practical applications 213–214
  support deflection 150
Crane mat life 108
Crane mats on slabs or pavement 201–202
Crane mats on structures 203–205
Crane mat stiffness 148–150
  flexible foundation 148
  rigid foundation 148
Crane mat strength and stiffness 87–128
  application summary 125–126
  dimensions and condition survey 87–92
    methodology 88–89
    survey results 89–92
  standards for timber properties 94–104
  timber defects 96–102
  timber species survey 92–94
Crane mats under crawler tracks 191–201
Crane mats under outriggers 190
Cranes supported on structures 78–83

applicable standards 82–83
barges 80–81, 201–205
bridges 80–81, 201–205
buildings 80–81, 201–205
framed structures 80–81
pavement and slabs on grade 79–80,
    201–205
Crane support loads 6–29, 184–185.
    See also Crawler track bearing pressures;
    See also Outrigger loads
Crane support requirements
    ASME B30.5 183–184
    OSHA (29 CFR 1926.1401) 182
    OSHA (29 CFR 1926.1402) 181–182
Crane support while traveling 205–207
CraniMax GmbH Crane Manager program 31
Crawler track bearing pressures 12–29
    blocked tracks 25–29
    calculation methods
        bases for the track bearing pressure
            calculation methods 22–23
        comparison of the calculation methods
            23–24
        equal LCG for both tracks 16–18
        equal rate of change for both tracks 18–22
    crawler track bearing pressure patterns
        31–32
    granular material over mats 24–25
    hard surface 12–15
    methods for reducing 24–29
    soft surface 12–15
    stepped distribution approximation 191–193
Cumaru 112

D

D9-12 94
D245-06(2011) 94–95, 96–104, 107, 116–117,
    119
D1196-12(2016) 49
D2555-17 94–95, 103, 111, 116–117, 120
Dabema 111
Dahoma 111
Defects in timbers 96–102
    checks 96–97, 98
    knots 97, 98
    pin holes 98
    pitch streaks 98
    pockets 98
    quality factor for modulus of elasticity 102
    shake 97, 99
    skips 99

slope of grain 99, 100
splits 99, 100
stain 99
torn grain 99
unsound wood 99
wane 99, 100–102
white speck 99
Demand/capacity ratio. *See* Utilization ratio
Density (soil) 52–53
Design factor – soil. *See* Factor of safety – soil
Design factor – steel 123
Design factor – timber. *See* Adjustment factor
Design values for other materials 124
Design values for timber mats 106–122
D-load 68–70
Duct banks 74
Ductile iron pipe 73
Duration of load 103, 107–110
    D245-06 107

E

Eccentrically loaded mats 151–154
Effective bearing area calculation method
    137–142
Effective bearing length 137–142
    bending strength 137–138
    deflection 139
    shear strength 138–139
"Effective Bearing Length of Crane Mats" 129
Ekki 111. *See* Azobe
Elastic properties of soil 48–52
    modulus of elasticity 51
    modulus of soil reaction 62–65
    modulus of subgrade reaction 49–51
    Poisson's ratio 51
Electrical duct banks 74
Embankments 78
Eucalyptus 111–115

F

Factor of safety – soil 46–47, 143, 186
Fiber-reinforced polymer (FRP) 124, 156
Flexible pipe 61
Frozen soil 53–54

G

Geotechnical exploration 41–45
    geotechnical engineer 41
    ground penetrating radar 45

magnetrometer 45
pocket penetrometer 42
visual examination 44–45
Grading requirements of non-U.S. standards 102
   Guyana Timber Grading Rules for
      Hardwoods 102
   Structural Timber – Visual Strength Grading
      of Tropical Hardwood 102
Grading requirements, visual 96–104
Greenheart 111
Ground bearing capacity 41–54
Ground penetrating radar 45
Grove Compu-Crane web site 30
Guidelines for the Design of Buried Steel Pipe
   60–62
Guyana Forestry Commission
   Guyana Timber Grading Rules for
      Hardwoods 102
Guyana rosewood 112. *See* Wamara
Guyana Timber Grading Rules for Hardwoods
   102

H

Hackberry 121

I

Inspection of crane mats 214–215

K

Knots 97, 98
Kobelco Ground Pressure/Outrigger Reaction
   Force web site 30

L

Laminated wood mats 176–177
Lateral earth pressure 76–77
Liebherr LICCON lift planner 18, 30
Life of a crane mat 108
Limit state stresses 118–122
Link-Belt Ground Bearing Information web
   site 30
Load duration. *See* Duration of load
Load duration factor 108–109, 118
Lower strength timber species 120–122
   cottonwood 121
   hackberry 121
   sweet gum 121
   sycamore 121

yellow poplar 121

M

M11 *Steel Pipe – A Guide for Design and*
   Installation 60
Magnetrometer 45
Manitowoc Ground Bearing Pressure Estimator
   program 18, 30, 184
Mat length based on ground bearing capacity
   130–133
Mat length based on mat strength 133–135
Mats continuous under both crawler tracks
   170–172
Mats of other materials 122–124, 154–161
   aluminum mats 124
   steel plates 158–161
   structural steel mats 122–123, 155–156
   synthetic mats 124, 155–157
Mixed hardwoods 92–93, 106–107
Mobile crane support design 181–218
Modern Sewer Design 70–73
Modulus of elasticity (soil) 51
Modulus of elasticity (wood) 110, 143
   quality factor 102
Modulus of rupture 111, 112, 121
Modulus of soil reaction 62–65
Modulus of subgrade reaction 49–51, 143–144
Moisture content 103
Mora 92–94, 95, 111–115
MS-1 *Thickness Design – Highways & Streets*
   82, 205
Multiple layers of mats – mats parallel 161–163,
   194–197
Multiple layers of mats – mats perpendicular
   163–170, 197–200

N

National Design Specification for Wood
   Construction 106–115, 118, 122
National Design Specification for Wood
   Construction, 1986 edition 106
National Lumber Grades Authority (NLGA)
   95–96, 107
NeLMA. *See* Northeastern Lumber
   Manufacturers Association (NeLMA)
NLGA. *See* National Lumber Grades Authority
Nonlinear beam-on-soil analysis 145–148
Northeastern Lumber Manufacturers Association
   (NeLMA) 95–96, 107
Northern Softwood Lumber Bureau (NSLB) 95

NSLB. *See* Northern Softwood Lumber Bureau

O

OSHA (29 CFR 1926.1401) 182
OSHA (29 CFR 1926.1402) 181–182
Outrigger loads 8–12
    calculation methods
        comparison of the calculation methods
            11–12
        geometric distribution 10–11
        rigid body 8–10
    front outriggers 12
    outrigger load patterns 32–33

P

Past practice of crane mat analysis 129–137
Percent utilization 136. *See also* Utilization
    ratio
Performance of the balanced mat analysis
    method 142–148
Plywood used under cranes 124–125
Pocket penetrometer 42
Poisson's ratio 51
Portland Cement Association
    Concrete Floors on Ground 82, 205
Posts and timbers 96, 98–99, 100, 106–107,
    114–117
Presumptive bearing pressures 47–48
Proposed strength and stiffness design values for
    timber mats 106–121
PVC pipe 73

Q

Quality factor for modulus of elasticity 102

R

Rain, effect on soil strength 47
Redwood Inspection Service (RIS) 96
Regulatory language 181–183
Responsibilities
    Crane Operator 183
    Lift Director 183
    Site Supervisor 183
Retaining walls and bulkheads 75–78
    lateral earth pressure 76–77
Rigid pipe 66
RIS. *See* Redwood Inspection Service

Rules for Building and Classing Steel Barges
    82, 205
Rules of thumb 206–213
    crawler cranes 207–208
    outrigger-supported cranes 208–213

S

Shake 97, 99
Size factor 103–104
Slope of grain 99, 100
Software for crane load calculations 29–31
    A1A Software, LLC 3D Lift Plan web site 31
    CraniMax GmbH Crane Manager program
        31
    Grove Compu-Crane web site 30
    Kobelco Ground Pressure/Outrigger Reaction
        Force web site 30
    Liebherr LICCON lift planner 30
    Link-Belt Ground Bearing Information web
        site 30
    Manitowoc Ground Bearing Pressure
        Estimator program 30
Soil bearing capacity 41, 45–47
    addressing low soil bearing capacity 54–56
    building code values 47–48
    consolidation 46
    factor of safety 46–47
    frozen soil 53–54
    immediate settlement 46
    presumptive bearing pressures 47–48
    rain, effect on soil strength 47
    Standard Penetration Test 47
    ultimate bearing capacity 46
Soil properties 45–54
    angle of internal friction 52–53
    density 52–53
    modulus of elasticity 51
    modulus of soil reaction 62–65
    modulus of subgrade reaction 49–51
    Poisson's ratio 51
Southern Pine Inspection Bureau (SPIB) 96
Specification for Aluminum Structures 124, 137
Specification for Structural Steel Buildings 77,
    82, 123, 137, 155, 172, 204
*Specification for Structural Steel Buildings –
    Allowable Stress Design and Plastic Design*
    123, 137, 155, 172
SPIB. *See* Southern Pine Inspection Bureau
Splits 99, 100
Standard Penetration Test 47
Standards for timber properties 94–104

adjustments 103–104
ASTM standards 94–95
defects 96–102
industry standards and guides 95–96
Steel mats 122–123
    allowable stresses 123
    crawler cranes 122
    outrigger-supported cranes 123
Steel plate over timber mats 172–176
Steel plates 158–161
Strength of buried pipes 60–73
    allowable bending stress (steel pipe) 62
    concrete pipe 65–70
    corrugated steel pipe 70–73
    ductile iron pipe 73
    flexible pipe 61
    PVC pipe 73
    rigid pipe 66
    steel pipe 60–65
    vitrified clay pipe 73
Strength of the supporting surface 41–86
    application summary 83
    geotechnical exploration 41–45
    soil properties 45–54
Strength ratio 96, 116–117
Structural steel mats 155–156
Structural Timber – Visual Strength Grading of Tropical Hardwood 102
Structures affected by surface loads 56–78
    buried pipes 60–73
    electrical duct banks 74
    embankments 78
    retaining walls and bulkheads 75–78
    utility vaults 73–74
Subsurface pipes 186–189
Subsurface structures 186
Suggested allowable stresses for timber crane mats 115
Support deflection 150
Supporting surface strength 186–189
    allowable ground bearing pressure 186
    allowable soil bearing pressure 186
    subsurface pipes 186–189
    subsurface structures 186
Support load patterns 31–33
Sweet gum 121
Sycamore 121
Synthetic mats 124, 155–158
Synthetic pad over timber mats 172–175

T

Three-edge-bearing load 70
Tie rods 104–106
Timber defects. *See* Defects in timbers
Timber mat dimensions and condition survey 87–92
    methodology 88–89
    survey results 89–92
Timber mat survey participants 126
Timber species survey 92–94
Tonka 112
Traveling, crane support while 205–207
Tropical hardwoods 111–117, 120–121
    azobe 92–94, 95, 111–115
    cumaru 112
    dabema 111
    dahoma 111–112
    ekki 111
    eucalyptus 111–115
    greenheart 111–114
    Guyana rosewood 112
    mora 92–93, 95, 111–115
    tonka 112–115
    wamara 112–114
Truncated pyramid 58–59, 67–68, 70

U

Ultimate bearing capacity 46, 143
Ultra high molecular weight (UHMW) plastic 124, 156
Underground utilities 73–74
    electrical duct banks 74
    utility vaults 73–74
Utilization ratio 135–136, 140, 142

V

Variable load effects 33–39
    change in boom tip load 34–36
    horizontal dynamic forces 36–39
    vertical impact 36
    wind load 38–39
Visual grading requirements 96–102
Vitrified clay pipe 73

W

Wamara 112–115
Wane 99, 100–102

WCLIB. *See* West Coast Lumber Inspection
        Bureau
West Coast Lumber Inspection Bureau (WCLIB)
        96, 107
Western Wood Products Association (WWPA)
        96
Wind load 38–39
*Wood Handbook – Wood as an Engineering
        Material* 95, 112–114
WWPA. *See* Western Wood Products
        Association

Y

Yellow poplar 121